EAA Series

EAA series is successor of the EAA Lecture Notes and supported by the European Actuarial Academy (EAA GmbH), founded on the 29 August, 2005 in Cologne (Germany) by the Actuarial Associations of Austria, Germany, the Netherlands and Switzerland. EAA offers actuarial education including examination, permanent education for certified actuaries and consulting on actuarial education.

actuarial-academy.com

More information about this series at http://www.springer.com/series/7879

Michael Radtke · Klaus D. Schmidt
Anja Schnaus
Editors

Handbook on Loss Reserving

 Springer

Editors
Michael Radtke
Lohmar
Germany

Anja Schnaus
Kerpen
Germany

Klaus D. Schmidt
Freital
Germany

Based on a translation from the German language edition: *Handbuch zur Schadenreservierung* by Michael Radtke and Klaus D. Schmidt Copyright © Verlag Versicherungswirtschaft GmbH 2012. All Rights Reserved.

ISSN 1869-6929   ISSN 1869-6937   (electronic)
EAA Series
ISBN 978-3-319-30054-2   ISBN 978-3-319-30056-6   (eBook)
DOI 10.1007/978-3-319-30056-6

Library of Congress Control Number: 2016942503

Mathematics Subject Classification (2010): 62-00

Printed on acid-free paper

This Springer imprint is published by Springer Nature
The registered company is Springer International Publishing AG Switzerland

Preface

In property and casualty insurance the provisions for payment obligations from losses that have occurred but have not yet been settled usually constitute the largest item on the liabilities side of an insurer's balance sheet. For this reason, the determination and evaluation of these technical provisions, which are also called *loss reserves*, is of considerable economic importance for every property and casualty insurer. Therefore, the application of actuarial methods of loss reserving is indispensable.

This *Handbook on Loss Reserving* presents the basic aspects of actuarial loss reserving. Besides the traditional methods it also includes a description of more recent ones and a discussion of certain problems occurring in actuarial practice, like inflation, scarce data, large claims, slow loss development, the use of market statistics, the need for simulation techniques, and last but not least, the task of calculating best estimates and ranges of future losses.

The actuarial methods of loss reserving form a substantial part of this book. These methods are presented in separate articles which are to a large extent self-contained. In the articles on traditional methods, the description of the method is accompanied by two numerical examples; these examples are the same for all methods and illustrate their sensitivity with respect to a small change in the data. While the traditional methods are univariate in the sense that they aim at prediction for a single portfolio of risks, the new multivariate methods, developed about ten years ago, aim at simultaneous prediction for several portfolios and take dependencies between these portfolios into account. Such methods are presented as well.

Almost all of the traditional methods are related to the *Bornhuetter–Ferguson principle*, which consists of an analytical part and a synthetical part. The analytical part provides a unified form of the predictors of most traditional methods such that the differences between these methods can be explained by the use of different estimators of parameters related to accident years or development years, and hence also by the use of different kinds of information, and the synthetical part consists of the construction of new methods by using new combinations of such estimators.

The methods of loss reserving and their properties can only be understood on the basis of stochastic models, which describe the generation of the run-off data and express the assumptions on the development (run-off) behaviour. For this reason, the articles on methods also discuss stochastic models that justify the respective method. By contrast, some other articles emphasize a stochastic model and then use the model together with a classical principle of mathematical statistics to construct a method of loss reserving.

There are basically two types of stochastic models that can be used to justify a method of loss reserving:

– *Development patterns* formalize the idea that, up to random fluctuations, the development of losses is identical for all different accident years, and they involve only assumptions on the expectations of the incremental or cumulative losses.
– *Linear models* and *credibility models* involve assumptions not only on the expectations but also on the variances and covariances of the incremental or cumulative losses. They thus enable the determination of the expected squared prediction error and its estimation.

While the traditional univariate methods result from heuristic considerations and were justified by a stochastic model later, the new multivariate methods result from generalizations of such models.

This book addresses actuarial students and academics as well as practicing actuaries. It is not intended as a complete presentation of all aspects of loss reserving, but rather as an invitation to gain an overview of the most important actuarial methods, to understand their underlying stochastic models and to get an idea of how to solve certain problems which may occur in practice. To proceed further and to become acquainted with other models and methods of loss reserving which are outside the scope of this book, the advanced reader may consult the survey articles by England & Verrall (2002) and by Schmidt (2012) and the monographs by Taylor (1986, 2000), and by Wüthrich & Merz (2008). We also refer to *A Bibliography on Loss Reserving*

http://www.math.tu-dresden.de/sto/schmidt/dsvm/reserve.pdf

which will be completed from time to time.

This *Handbook on Loss Reserving* is a free translation of the second edition of the *Handbuch zur Schadenreservierung*, published in 2012 as an update and extension of its first edition which appeared in 2004. A few articles of the German editions have been excluded since they are either outdated or specific to the German market.

The articles of this book are arranged in alphabetical order. They allow for a quick access to the different subjects, and the following guide *How to Read This Book* contains several hints on connections between certain articles and on possible starting points for reading this book.

The editors are most grateful to the authors who contributed to this book and whose expertise enabled a concise and consistent presentation of diverse theoretical and practical aspects of loss reserving, to Helga Mettke and Christiane Weber for their very delicious work in producing the graphics of this book, and to Christiane Weber for thoughtful proofreading.

Lohmar, Germany   Michael Radtke
Freital, Germany   Klaus D. Schmidt
Kerpen, Germany   Anja Schnaus
December 2015

How to Read This Book

The articles of this book are to a large extent self-contained. Nonetheless, there are many interconnections between the different articles, and these are accessible via the keywords given at the end of an article.

With regard to actuarial practice, many articles focus on methods. In spite of this there is a close relationship between methods, stochastic models, and general principles of statistics. For example, many methods can be justified by the assumption of an underlying development pattern, which presents an elementary stochastic model, and for some methods it is even possible to show that the pre-dictors of future losses are optimal in a certain sense since they turn out to be the Gauss–Markov predictors in a suitable linear model.

For a general background and a systematic approach to methods and models of loss reserving, and to become familiar with the notation, it is useful to start with the articles *Run-Off Data* and *Run-Off Triangles* as well as *Development Patterns* (*Basics*), *Development Patterns* (*Estimation*) and *Loss Ratios*:

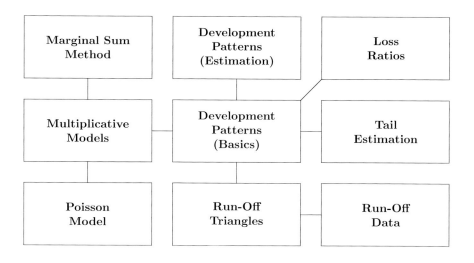

A central group of articles presents the basic methods of loss reserving and the *Bornhuetter–Ferguson principle*, which provides a general framework for a unified presentation and possible extensions of these methods:

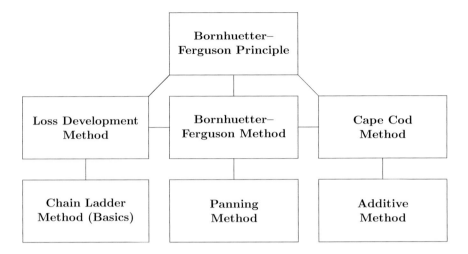

Some of these methods can also be justified by a linear model or can be extended to the multivariate case involving different lines of business or different kinds of data of the same line of business:

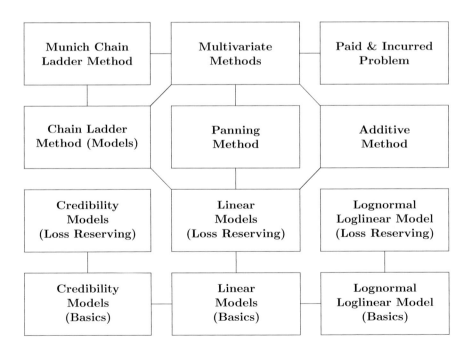

Of course, there are many other possible paths through the variety of methods and models of loss reserving.

The notions and also the notation used in loss reserving are far from being uniform in the literature. In this book we have tried to use uniform notions and notation as far as possible in order to simplify the recognition of interrelations between the different topics. The subject index should be helpful in this regard.

We have also tried not to burden the book too much with technicalities. For example, in an identity like

$$a = b/c$$

it is tacitly assumed that $c \neq 0$. This remark is far from being trivial since many methods of loss reserving involve divisions and are thus not applicable when the data lead to a division by 0.

Contents

Additive Method

Klaus D. Schmidt and Mathias Zocher

Consider the run-off square of incremental losses:

Accident year	Development year						
	0	1	... k	... $n-i$	... $n-1$	n	
0	$Z_{0,0}$	$Z_{0,1}$	... $Z_{0,k}$	... $Z_{0,n-i}$	... $Z_{0,n-1}$	$Z_{0,n}$	
1	$Z_{1,0}$	$Z_{1,1}$	... $Z_{1,k}$	... $Z_{1,n-i}$	... $Z_{1,n-1}$	$Z_{1,n}$	
⋮	⋮	⋮	⋮	⋮	⋮	⋮	
i	$Z_{i,0}$	$Z_{i,1}$	... $Z_{i,k}$	... $Z_{i,n-i}$	... $Z_{i,n-1}$	$Z_{i,n}$	
⋮	⋮	⋮	⋮	⋮	⋮	⋮	
$n-k$	$Z_{n-k,0}$	$Z_{n-k,1}$	... $Z_{n-k,k}$	... $Z_{n-k,n-i}$	... $Z_{n-k,n-1}$	$Z_{n-k,n}$	
⋮	⋮	⋮	⋮	⋮	⋮	⋮	
$n-1$	$Z_{n-1,0}$	$Z_{n-1,1}$	... $Z_{n-1,k}$	... $Z_{n-1,n-i}$	... $Z_{n-1,n-1}$	$Z_{n-1,n}$	
n	$Z_{n,0}$	$Z_{n,1}$	... $Z_{n,k}$	... $Z_{n,n-i}$	... $Z_{n,n-1}$	$Z_{n,n}$	

We assume that the incremental losses $Z_{i,k}$ are observable for $i + k \leq n$ and that they are non-observable for $i + k \geq n + 1$. For $i, k \in \{0, 1, \ldots, n\}$ we denote by

$$S_{i,k} := \sum_{l=0}^{k} Z_{i,l}$$

the cumulative loss from accident year i in development year k.

K.D. Schmidt (✉) · M. Zocher
Technische Universität Dresden, Dresden, Germany
e-mail: klaus.d.schmidt@tu-dresden.de

© Springer International Publishing Switzerland 2016
M. Radtke et al. (eds.), *Handbook on Loss Reserving*,
EAA Series, DOI 10.1007/978-3-319-30056-6_1

The additive method, which is also called the *incremental loss ratio method*, involves known *volume measures* $v_0, v_1, \ldots, v_n$ of the accident years and is based on the development pattern for incremental loss ratios:

Development Pattern for Incremental Loss Ratios: *There exist parameters* $\zeta_0, \zeta_1, \ldots, \zeta_n$ *such that the identity*

$$E\left[\frac{Z_{i,k}}{v_i}\right] = \frac{E[Z_{i,k}]}{v_i} = \zeta_k$$

holds for all $k \in \{0, 1, \ldots, n\}$ *and for all* $i \in \{0, 1, \ldots, n\}$.

In this article, we assume that there exists a development pattern for incremental loss ratios. Then the parameters $\vartheta_0, \vartheta_1, \ldots, \vartheta_n$ given by

$$\vartheta_k := \frac{\zeta_k}{\sum_{l=0}^{n} \zeta_l}$$

form a *development pattern for incremental quotas* and the parameters $\gamma_0, \gamma_1, \ldots, \gamma_n$ given by

$$\gamma_k := \frac{\sum_{l=0}^{k} \zeta_l}{\sum_{l=0}^{n} \zeta_l}$$

form a *development pattern for quotas*. Moreover, the identity

$$E[Z_{i,k}] = v_i \zeta_k$$

yields the existence of a *multiplicative model*, and the identity

$$E\left[\frac{S_{i,n}}{v_i}\right] = \sum_{l=0}^{n} E\left[\frac{Z_{i,l}}{v_i}\right] = \sum_{l=0}^{n} \zeta_l$$

shows that the expected ultimate loss ratios of all accident years are identical.

The additive method consists of two steps:

- For every development year $k \in \{0, 1, \ldots, n\}$, the expected incremental loss ratio ζ_k is estimated by the *additive incremental loss ratio*

$$\zeta_k^{\mathrm{AD}} := \frac{\sum_{j=0}^{n-k} Z_{j,k}}{\sum_{j=0}^{n-k} v_j}$$

Since

$$\zeta_k^{\mathrm{AD}} = \sum_{j=0}^{n-k} \frac{v_j}{\sum_{h=0}^{n-k} v_h} \frac{Z_{j,k}}{v_j}$$

the additive incremental loss ratio ζ_k^{AD} is a weighted mean of the observable *individual incremental loss ratios* $Z_{j,k}/v_j$ of development year k, with weights proportional to the volume measures of the accident years.
- For every accident year i and every development year k such that $i + k \geq n + 1$, the future incremental loss $Z_{i,k}$ is predicted by the *additive predictor*

$$Z_{i,k}^{AD} := v_i \zeta_k^{AD}$$

The definition of the additive predictors of the incremental losses reflects the identity

$$E[Z_{i,k}] = v_i \zeta_k$$

which results from the development pattern for incremental loss ratios.

Using the additive predictors of the future incremental losses, we define the *additive predictors*

$$S_{i,k}^{AD} := S_{i,n-i} + \sum_{l=n-i+1}^{k} Z_{i,l}^{AD} = S_{i,n-i} + v_i \sum_{l=n-i+1}^{k} \zeta_l^{AD}$$

of the future cumulative losses $S_{i,k}$ and the *additive predictors*

$$R_i^{AD} := \sum_{l=n-i+1}^{n} Z_{i,l}^{AD}$$

$$R_{(c)}^{AD} := \sum_{l=c-n}^{n} Z_{c-l,l}^{AD}$$

$$R^{AD} := \sum_{l=1}^{n} \sum_{j=n-l+1}^{n} Z_{j,l}^{AD}$$

of the accident year reserves R_i with $i \in \{1, \ldots, n\}$, the calender year reserves $R_{(c)}$ with $c \in \{n+1, \ldots, 2n\}$ and the aggregate loss reserve R. The additive predictors of reserves are also called *additive reserves*. Moreover, the *additive ultimate loss ratio*

$$\kappa^{AD} := \sum_{l=0}^{n} \zeta_l^{AD}$$

is an estimator of the expected ultimate loss ratio

$$\kappa := E\left[\frac{S_{i,n}}{v_i}\right] = \sum_{l=0}^{n} E\left[\frac{Z_{i,l}}{v_i}\right] = \sum_{l=0}^{n} \zeta_l$$

which is identical for all accident years.

Example A. Calculation of the additive predictors of incremental losses:

Accident	Development year k						Volume	Sum
year i	0	1	2	3	4	5	v_i	
0	1001	854	568	565	347	148	4025	
1	1113	990	671	648	422	164	4456	
2	1265	1168	800	744	482	195	5315	
3	1490	1383	1007	849	543	220	5986	
4	1725	1536	1068	984	629	255	6939	
5	1889	1811	1256	1157	740	300	8158	
ζ_k^{AD}	0.24	0.22	0.15	0.14	0.09	0.04		0.89
ϑ_k^{AD}	0.27	0.25	0.17	0.16	0.10	0.04		1
γ_k^{AD}	0.27	0.52	0.70	0.86	0.96	1		

Reserves:

Accident year i	Reserve R_i^{AD}
1	164
2	677
3	1612
4	2937
5	5264
total	10654

Calendar year c	Reserve $R_{(c)}^{AD}$
6	4374
7	2979
8	2007
9	995
10	300
total	10654

The estimators of the development pattern for incremental quotas and quotas are not needed for the additive method and are given only for the sake of comparison with other methods.

Example B. In this example the incremental loss $Z_{4,1}$ is increased by 1000:

Accident	Development year k						Volume	Sum
year i	0	1	2	3	4	5	v_i	
0	1001	854	568	565	347	148	4025	
1	1113	990	671	648	422	164	4456	
2	1265	1168	800	744	482	195	5315	
3	1490	1383	1007	849	543	220	5986	
4	1725	2536	1068	984	629	255	6939	
5	1889	2116	1256	1157	740	300	8158	
ζ_k^{AD}	0.24	0.26	0.15	0.14	0.09	0.04		0.93
ϑ_k^{AD}	0.26	0.28	0.17	0.15	0.10	0.04		1
γ_k^{AD}	0.26	0.54	0.71	0.86	0.96	1		

Reserves:

Accident year i	Reserve R_i^{AD}
1	164
2	677
3	1612
4	2937
5	5569
total	10959

Calendar year c	Reserve $R_{(c)}^{\mathrm{AD}}$
6	4679
7	2979
8	2007
9	995
10	300
total	10959

The outlier $Z_{4,1}$ affects the estimator of the parameter ζ_1 and hence the predictors of the incremental loss $Z_{5,1}$, the cumulative losses $S_{5,k}$ with $k \in \{1, \ldots, 5\}$, the accident year reserve R_5 and the calendar year reserve $R_{(6)}$.

Bornhuetter–Ferguson Principle

Define now

$$\gamma_k^{\mathrm{AD}} := \frac{\sum_{l=0}^{k} \zeta_l^{\mathrm{AD}}}{\sum_{l=0}^{n} \zeta_l^{\mathrm{AD}}} \quad \text{and} \quad \alpha_i^{\mathrm{AD}} := v_i \sum_{l=0}^{n} \zeta_l^{\mathrm{AD}}$$

Then the additive predictors of the future cumulative losses satisfy

$$S_{i,k}^{\mathrm{AD}} = S_{i,n-i} + \left(\gamma_k^{\mathrm{AD}} - \gamma_{n-i}^{\mathrm{AD}}\right) \alpha_i^{\mathrm{AD}}$$

Therefore, the additive method is subject to the *Bornhuetter–Ferguson principle*. Because of the definition of κ^{AD}, we also have $\alpha_i^{\mathrm{AD}} = v_i \kappa^{\mathrm{AD}}$ and hence

$$S_{i,k}^{\mathrm{AD}} = S_{i,n-i} + \left(\gamma_k^{\mathrm{AD}} - \gamma_{n-i}^{\mathrm{AD}}\right) v_i \kappa^{\mathrm{AD}}$$

Moreover, if the Cape Cod ultimate loss ratio κ^{CC} is computed by using the *additive quotas* γ_k^{AD}, then it satisfies

$$\kappa^{\mathrm{CC}} = \kappa^{\mathrm{AD}}$$

This means that the additive method is a special case of the *Cape Cod method*. Furthermore, since the development pattern for incremental loss ratios yields a development pattern $\gamma_0, \gamma_1, \ldots, \gamma_n$ for quotas, we have

$$E\left[\frac{S_{i,k}}{v_i \gamma_k}\right] = E\left[\frac{S_{i,n}}{v_i}\right] = \kappa$$

for all $i, k \in \{0, 1, \ldots, n\}$. This is an assumption of the Cape Cod model.

Linear Model

The development pattern for incremental loss ratios concerns the structure of the expectations of the incremental losses. This elementary model can be refined by adding an assumption on the structure of their covariances. Such an assumption is part of the *additive model*:

Additive Model: *There exist known volume measures* $v_0, v_1, \ldots, v_n$ *of the accident years as well as unknown parameters* $\zeta_0, \zeta_1, \ldots, \zeta_n$ *and parameters* $\sigma_0^2, \sigma_1^2, \ldots, \sigma_n^2$ *such that the identities*

$$E\left[\frac{Z_{i,k}}{v_i}\right] = \zeta_k$$

$$\mathrm{cov}\left[\frac{Z_{i,k}}{v_i}, \frac{Z_{j,l}}{v_j}\right] = \frac{1}{v_i} \sigma_k^2 \, \delta_{i,j} \, \delta_{k,l}$$

hold for all $i, j, k, l \in \{0, 1, \ldots, n\}$.

The conditions of the additive model can be also represented in the form

$$E[Z_{i,k}] = v_i \, \zeta_k$$
$$\mathrm{cov}[Z_{i,k}, Z_{j,l}] = v_i \, \sigma_k^2 \, \delta_{i,j} \, \delta_{k,l}$$

Therefore, the additive model is a *linear model* and it is obvious that all additive predictors are linear in the observable incremental losses. Further properties of the additive predictors result from the theory of linear models:

Theorem. *In the additive model, the additive predictor of the future incremental loss* $Z_{i,k}$ *is unbiased, it is optimal in the sense that it minimizes the expected squared prediction error*

$$E\left[\left(\widehat{Z}_{i,k} - Z_{i,k}\right)^2\right]$$

over all unbiased linear predictors $\widehat{Z}_{i,k}$ *of* $Z_{i,k}$, *and it is the only predictor having this property. These properties also hold for the additive predictors of cumulative losses and reserves.*

Under the assumptions of the additive model, the theorem asserts that the additive predictors are precisely the *Gauss–Markov predictors*. In particular, it is possible to determine the expected squared prediction errors of the additive reserves and one obtains

$$E\left[\left(R_i^{\mathrm{AD}} - R_i\right)^2\right] = v_i^2 \sum_{l=n-i+1}^{n} \left(\frac{1}{\sum_{h=0}^{n-l} v_h} + \frac{1}{v_i}\right) \sigma_l^2$$

$$E\left[\left(R_{(c)}^{\mathrm{AD}} - R_{(c)}\right)^2\right] = \sum_{l=c-n}^{n} v_{c-l}^2 \left(\frac{1}{\sum_{h=0}^{n-l} v_h} + \frac{1}{v_{c-l}}\right) \sigma_l^2$$

$$E\left[\left(R^{\mathrm{AD}} - R\right)^2\right] = \sum_{l=1}^{n} \left(\sum_{j=n-l+1}^{n} v_j\right)^2 \left(\frac{1}{\sum_{h=0}^{n-l} v_h} + \frac{1}{\sum_{h=n-l+1}^{n} v_h}\right) \sigma_l^2$$

To estimate the prediction errors, one has to replace the variance parameters $\sigma_1^2, \ldots, \sigma_n^2$ occurring in these formulae by appropriate estimators. Usually, the unbiased estimators

$$\widehat{\sigma}_k^2 := \frac{1}{n-k} \sum_{j=0}^{n-k} v_j \left(\frac{Z_{j,k}}{v_j} - \zeta_k^{\mathrm{AD}}\right)^2$$

are chosen for $k \in \{1, \ldots, n-1\}$, and an estimator $\widehat{\sigma}_n^2$ is determined by extrapolation.

Remarks

The structure of the additive method is very similar to that of the *chain ladder method* and that of the *Panning method*. Correspondingly, the additive model is quite similar to the *chain ladder model of Schnaus* and the *Panning model*.

The additive method can be modified by changing the weights in the additive incremental loss ratios

$$\zeta_k^{\mathrm{AD}} = \sum_{j=0}^{n-k} \frac{v_j}{\sum_{h=0}^{n-k} v_h} \frac{Z_{j,k}}{v_j}$$

and such a change of the weights can be captured by an appropriate change of the accident year factors $1/v_i$ in the covariance condition

$$\mathrm{cov}\left[\frac{Z_{i,k}}{v_i}, \frac{Z_{j,l}}{v_j}\right] = \frac{1}{v_i} \sigma_k^2 \delta_{i,j} \delta_{k,l}$$

of the additive model.

It is interesting to note that there is also a micro model leading to the additive model:

- Assume that the volume measures are positive integers.

- Assume further that for every cell (i, k) with $i, k \in \{0, 1, \ldots, n\}$ there exists a family of random variables $\{X_{i,k,l}\}_{l \in \{1, \ldots, v_i\}}$ with $E[X_{i,k,l}] = \zeta_k$ and $\text{var}[X_{i,k,l}] = \sigma_k^2$ as well as

$$Z_{i,k} = \sum_{l=1}^{v_i} X_{i,k,l}$$

- Assume also that any two of the random variables $X_{i,k,l}$ are uncorrelated.

Then the family $\{Z_{i,k}\}_{i,k \in \{0,1,\ldots,n\}}$ satisfies the assumptions of the additive model. The quantities of this micro model may be interpreted as follows: In accident year i there are v_i contracts, and for contract $l \in \{1, \ldots, v_i\}$ from accident year i the incremental loss in development year k is given by $X_{i,k,l}$.

Notes

Keywords: Aggregation, Bornhuetter–Ferguson Method, Bornhuetter–Ferguson Principle, Cape Cod Method, Chain Ladder Method (Basics), Development Pattern (Basics), Development Pattern (Estimation), Linear Models (Loss Reserving), Loss Ratios, Multiplicative Models, Multivariate Methods, Paid & Incurred Problem, Panning Method, Run-Off Triangles, Volume Measures.

References: Ludwig, Schmeißer & Thänert [2009], Mack [2002], Schmidt [2009, 2012], Schmidt & Zocher [2008].

Aggregation

Sebastian Fuchs, Heinz J. Klemmt and Klaus D. Schmidt

For small or highly volatile portfolios the standard methods of loss reserving tend to produce highly volatile predictors of future losses and hence of reserves. One might be tempted to combine a small or highly volatile portfolio with a large and stable one and to apply the corresponding methods to the resulting total portfolio. A typical example of such a situation is the combination of bodily injury claims with pure property damage claims in *motor third party liability insurance* insurance.

However, aggregation of sub-portfolios to a total portfolio turns out to be problematic since it can lead to a systematic distortion of the predictors. This is, in particular, the case when the sub-portfolios show different development patterns and develop differently also over the accident years. Moreover, if the standard methods of loss reserving are interpreted not just as algorithms but rather as statistical methods based on a stochastic model, then the problem arises that a model which is acceptable for each of the sub-portfolios will not necessarily be appropriate for the total portfolio. We discuss these aspects of aggregation for the chain ladder method and the additive method.

S. Fuchs · H.J. Klemmt · K.D. Schmidt (✉)
Technische Universität Dresden, Dresden, Germany
e-mail: klaus.d.schmidt@tu-dresden.de

© Springer International Publishing Switzerland 2016   9
M. Radtke et al. (eds.), *Handbook on Loss Reserving*,
EAA Series, DOI 10.1007/978-3-319-30056-6_2

Consider the run-off square of incremental losses:

Accident	Development year							
year	0	1	... k	... $n-i$	... $n-1$	n		
0	$Z_{0,0}$	$Z_{0,1}$	... $Z_{0,k}$	... $Z_{0,n-i}$	... $Z_{0,n-1}$	$Z_{0,n}$		
1	$Z_{1,0}$	$Z_{1,1}$	... $Z_{1,k}$	... $Z_{1,n-i}$	... $Z_{1,n-1}$	$Z_{1,n}$		
$\vdots$	$\vdots$	$\vdots$	$\vdots$	$\vdots$	$\vdots$	$\vdots$		
i	$Z_{i,0}$	$Z_{i,1}$	... $Z_{i,k}$	... $Z_{i,n-i}$	... $Z_{i,n-1}$	$Z_{i,n}$		
$\vdots$	$\vdots$	$\vdots$	$\vdots$	$\vdots$	$\vdots$	$\vdots$		
$n-k$	$Z_{n-k,0}$	$Z_{n-k,1}$	... $Z_{n-k,k}$	... $Z_{n-k,n-i}$	... $Z_{n-k,n-1}$	$Z_{n-k,n}$		
$\vdots$	$\vdots$	$\vdots$	$\vdots$	$\vdots$	$\vdots$	$\vdots$		
$n-1$	$Z_{n-1,0}$	$Z_{n-1,1}$	... $Z_{n-1,k}$	... $Z_{n-1,n-i}$	... $Z_{n-1,n-1}$	$Z_{n-1,n}$		
n	$Z_{n,0}$	$Z_{n,1}$	... $Z_{n,k}$	... $Z_{n,n-i}$	... $Z_{n,n-1}$	$Z_{n,n}$		

We assume that the incremental losses $Z_{i,k}$ are observable for $i + k \leq n$ and that they are non-observable for $i + k \geq n + 1$. For $i, k \in \{0, 1, \ldots, n\}$, let

$$S_{i,k} := \sum_{l=0}^{k} Z_{i,l}$$

denote the cumulative loss from accident year i in development year k.

Chain Ladder Method

The chain ladder method is usually described by means of the cumulative losses. It is based on the *chain ladder factors*

$$\varphi_k^{\mathrm{CL}} := \frac{\sum_{j=0}^{n-k} S_{j,k}}{\sum_{j=0}^{n-k} S_{j,k-1}}$$

with $k \in \{1, \ldots, n\}$ and it consists primarily in the prediction of the future cumulative losses $S_{i,k}$ with $i + k \geq n + 1$ by the *chain ladder predictors*

$$S_{i,k}^{\mathrm{CL}} := S_{i,n-i} \prod_{l=n-i+1}^{k} \varphi_l^{\mathrm{CL}}$$

For the prediction of the future incremental losses $Z_{i,k}$ with $i + k \geq n + 1$ one uses the *chain ladder predictors*

$$Z_{i,k}^{\mathrm{CL}} := S_{i,n-i} (\varphi_k^{\mathrm{CL}} - 1) \prod_{l=n-i+1}^{k-1} \varphi_l^{\mathrm{CL}}$$

(with $Z_{i,n-i+1}^{\mathrm{CL}} = S_{i,n-i}\,(\varphi_{n-i+1}^{\mathrm{CL}} - 1)$) from which the *chain ladder predictors* of the reserves result by summation.

By analogy with the chain ladder factors, we define for $i \in \{1, \ldots, n\}$ the *dual chain ladder factors*

$$\psi_i^{\mathrm{CL}} := \frac{\sum_{j=0}^{i} S_{j,n-i}}{\sum_{j=0}^{i-1} S_{j,n-i}}$$

Here the analogy and the notion of duality result from the identities

$$\varphi_k^{\mathrm{CL}} = \frac{\sum_{j=0}^{n-k} \sum_{l=0}^{k} Z_{j,l}}{\sum_{j=0}^{n-k} \sum_{l=0}^{k-1} Z_{j,l}} \quad \text{and} \quad \psi_i^{\mathrm{CL}} = \frac{\sum_{l=0}^{n-i} \sum_{j=0}^{i} Z_{j,l}}{\sum_{l=0}^{n-i} \sum_{j=0}^{i-1} Z_{j,l}}$$

The dual chain ladder factors are exactly the chain ladder factor in the *reflected run-off triangle* of incremental losses, in which the roles of accident years and of development years are interchanged. Therefore they describe the development over accident years instead of development years.

We consider now two sub-portfolios with the respective incremental losses $\bar{Z}_{i,k} > 0$ and $\tilde{Z}_{i,k} > 0$ as well as the total portfolio with the incremental losses $Z_{i,k} := \bar{Z}_{i,k} + \tilde{Z}_{i,k}$. We also denote all other quantities of the sub-portfolios in the same way as the incremental losses.

Theorem.

(1) If $\bar{\varphi}_k^{\mathrm{CL}} > \tilde{\varphi}_k^{\mathrm{CL}}$ and $\bar{\psi}_i^{\mathrm{CL}} > \tilde{\psi}_i^{\mathrm{CL}}$ holds for all $i, k \in \{1, \ldots, n\}$, then the inequality

$$\bar{Z}_{i,k}^{\mathrm{CL}} + \tilde{Z}_{i,k}^{\mathrm{CL}} > Z_{i,k}^{\mathrm{CL}}$$

holds for all $i, k \in \{1, \ldots, n\}$ such that $i + k \geq n + 1$.

(2) If $\bar{\varphi}_k^{\mathrm{CL}} = \tilde{\varphi}_k^{\mathrm{CL}}$ holds for all $i, k \in \{1, \ldots, n\}$, then the identity

$$\bar{Z}_{i,k}^{\mathrm{CL}} + \tilde{Z}_{i,k}^{\mathrm{CL}} = Z_{i,k}^{\mathrm{CL}}$$

holds for all $i, k \in \{1, \ldots, n\}$ such that $i + k \geq n + 1$.

(3) If $\bar{\varphi}_k^{\mathrm{CL}} < \tilde{\varphi}_k^{\mathrm{CL}}$ and $\bar{\psi}_i^{\mathrm{CL}} > \tilde{\psi}_i^{\mathrm{CL}}$ holds for all $i, k \in \{1, \ldots, n\}$, then the inequality

$$\bar{Z}_{i,k}^{\mathrm{CL}} + \tilde{Z}_{i,k}^{\mathrm{CL}} < Z_{i,k}^{\mathrm{CL}}$$

holds for all $i, k \in \{1, \ldots, n\}$ such that $i + k \geq n + 1$.

By summation, the results of the theorem for the chain ladder predictors of incremental losses yield corresponding results for the chain ladder predictors of cumulative losses and for the chain ladder reserves.

Example. Sub-portfolio I: Incremental losses and predictors of incremental losses:

Accident year i	Development year k				$\bar{\psi}_i^{\mathrm{CL}}$
	0	1	2	3	
0	230	110	60	20	
1	240	120	80	22	2.10
2	230	120	70	21	1.50
3	280	140	84	25	1.40
$\bar{\varphi}_k^{\mathrm{CL}}$		1.50	1.20	1.05	

Sub-portfolio II: Incremental losses and predictors of incremental losses:

Accident year i	Development year k				$\bar{\psi}_i^{\mathrm{CL}}$
	0	1	2	3	
0	780	140	80	10	
1	760	120	100	10	1.98
2	410	130	54	6	1.30
3	390	78	47	5	1.20
$\bar{\varphi}_k^{\mathrm{CL}}$		1.20	1.10	1.01	

Sums of the predictors of the two sub-portfolios:

Accident year i	Development year k				
	0	1	2	3	
0					
1				32	
2			124	27	
3		218	131	30	

Total portfolio: Incremental losses and predictors of incremental losses:

Accident year i	Development year k				
	0	1	2	3	
0	1010	250	140	30	
1	1000	240	180	30	
2	640	250	114	21	
3	670	187	110	21	
φ_k^{CL}		1.28	1.13	1.02	

The results confirm assertion (1) of the theorem.

The chain ladder method is based on the assumption of the existence of a development pattern for factors.

- If the existence of a development pattern for factors is assumed for each of the sub-portfolios, then there exist parameters $\bar{\varphi}_k$ and $\tilde{\varphi}_k$ such that

$$E[\bar{S}_{i,k}] = E[\bar{S}_{i,k-1}]\,\bar{\varphi}_k$$
$$E[\tilde{S}_{i,k}] = E[\tilde{S}_{i,k-1}]\,\tilde{\varphi}_k$$

  holds for all $k \in \{1, \ldots, n\}$ and $i \in \{0, 1, \ldots, n\}$.

- If the existence of a development pattern for factors is assumed for the total portfolio, then there exist parameters φ_k such that

$$E[S_{i,k}] = E[S_{i,k-1}]\,\varphi_k$$

  holds for all $k \in \{1, \ldots, n\}$ and $i \in \{0, 1, \ldots, n\}$.

It thus follows that a development pattern for factors exists for each of the sub-portfolios and also for the total portfolio if and only if there exists, for every $k \in \{1, \ldots, n\}$, some c_{k-1} such that the identity

$$\frac{E[\bar{S}_{i,k-1}]}{E[\tilde{S}_{i,k-1}]} = c_{k-1}$$

holds for all $i \in \{0, 1, \ldots, n\}$. As this proportionality condition is not plausible in general, this raises the problem of a consistent modelling of the sub-portfolios and the total portfolio.

One possibility of a consistent modelling of the sub-portfolios and the total portfolio is provided by the *multivariate chain ladder model*, which provides a justification of the *multivariate chain ladder method*. The multivariate chain ladder model describes not only the individual sub-portfolios, but also the correlations between the sub-portfolios.

In actuarial practice, the application of the multivariate chain ladder method may cause problems, but the method represents a *benchmark* and in many cases the multivariate chain ladder predictors are approximated quite well by the univariate chain ladder predictors for the individual sub-portfolios.

Additive Method

The additive method uses known *volume measures* $v_0, v_1, \ldots, v_n$ of the accident years. It is based on the *additive incremental loss ratios*

$$\zeta_k^{\mathrm{AD}} := \frac{\sum_{j=0}^{n-k} Z_{j,k}}{\sum_{j=0}^{n-k} v_j}$$

with $k \in \{0, 1, \dots, n\}$ and it consists primarily in the prediction of the future incremental losses $Z_{i,k}$ with $i + k \geq n + 1$ by the *additive predictors*

$$Z_{i,k}^{\mathrm{AD}} := v_i \zeta_k^{\mathrm{AD}}$$

from which the *additive predictors* of the future cumulative losses and of the reserves result by summation.

We consider now two sub-portfolios with the respective incremental losses $\bar{Z}_{i,k} > 0$ and $\tilde{Z}_{i,k} > 0$ and the respective volume measures $\bar{v}_i > 0$ and $\tilde{v}_i > 0$ as well as the total portfolio with the incremental losses $Z_{i,k} := \bar{Z}_{i,k} + \tilde{Z}_{i,k}$ and the volume measures $v_i := \bar{v}_i + \tilde{v}_i$. We also denote all other quantities of the sub-portfolios in the same way as the incremental losses and the volume measures.

Lemma. *For all $i, k \in \{1, \dots, n\}$ such that $i + k \geq n + 1$ there exists a constant $v_{i,k} > 0$ determined by the volume measures such that*

$$\bar{Z}_{i,k}^{\mathrm{AD}} + \tilde{Z}_{i,k}^{\mathrm{AD}} - Z_{i,k}^{\mathrm{AD}} = v_{i,k} \left(\frac{\bar{v}_i}{\sum_{j=0}^{n-k} \bar{v}_j} - \frac{\tilde{v}_i}{\sum_{j=0}^{n-k} \tilde{v}_j} \right) \left(\bar{\zeta}_k^{\mathrm{AD}} - \tilde{\zeta}_k^{\mathrm{AD}} \right)$$

This lemma provides a complete solution to the problem of additivity for the additive method (and even for the additive predictors of the individual future incremental losses). In particular, the additive method is always additive if there exists some c such that the identity

$$\bar{v}_i / \tilde{v}_i = c$$

holds for all $i \in \{0, 1, \dots, n\}$.

An analogon to the theorem on the additivity in the chain ladder method results immediately from the lemma:

Theorem.

(1) *If $\bar{\zeta}_k^{\mathrm{AD}} > \tilde{\zeta}_k^{\mathrm{AD}}$ and $\bar{v}_i / \sum_{j=0}^{n-k} \bar{v}_j > \tilde{v}_i / \sum_{j=0}^{n-k} \tilde{v}_j$ holds for all $i, k \in \{1, \dots, n\}$ such that $i + k \geq n + 1$, then the inequality*

$$\bar{Z}_{i,k}^{\mathrm{AD}} + \tilde{Z}_{i,k}^{\mathrm{AD}} > Z_{i,k}^{\mathrm{AD}}$$

holds for all $i, k \in \{1, \dots, n\}$ such that $i + k \geq n + 1$.

(2) *If $\bar{\zeta}_k^{\mathrm{AD}} = \tilde{\zeta}_k^{\mathrm{AD}}$ or $\bar{v}_i / \sum_{j=0}^{n-k} \bar{v}_j = \tilde{v}_i / \sum_{j=0}^{n-k} \tilde{v}_j$ holds for all $i, k \in \{1, \dots, n\}$ such that $i + k \geq n + 1$, then the identity*

$$\bar{Z}_{i,k}^{\mathrm{AD}} + \tilde{Z}_{i,k}^{\mathrm{AD}} = Z_{i,k}^{\mathrm{AD}}$$

holds for all $i, k \in \{1, \dots, n\}$ such that $i + k \geq n + 1$.

(3) *If $\bar{\zeta}_k^{AD} < \tilde{\zeta}_k^{AD}$ and $\bar{v}_i / \sum_{j=0}^{n-k} \bar{v}_j > \tilde{v}_i / \sum_{j=0}^{n-k} \tilde{v}_j$ holds for all $i, k \in \{1, \ldots, n\}$ such that $i + k \geq n + 1$, then the inequality*

$$\bar{Z}_{i,k}^{AD} + \tilde{Z}_{i,k}^{AD} < Z_{i,k}^{AD}$$

holds for all $i, k \in \{1, \ldots, n\}$ such that $i + k \geq n + 1$.

By summation, the results of the theorem for the additive predictors of incremental losses yield corresponding results for the additive predictors of cumulative losses and for the additive reserves.

The additive method is based on the assumption of the existence of a development pattern for incremental loss ratios.

- If the existence of a development pattern for incremental loss ratios is assumed for each of the sub-portfolios, then there exist parameters $\bar{\zeta}_k$ and $\tilde{\zeta}_k$ such that

$$E[\bar{Z}_{i,k}] = \bar{v}_i \bar{\zeta}_k$$
$$E[\tilde{Z}_{i,k}] = \tilde{v}_i \tilde{\zeta}_k$$

  holds for all $k \in \{0, 1, \ldots, n\}$ and $i \in \{0, 1, \ldots, n\}$.
- If the existence of a development pattern for incremental loss ratios is assumed for the total portfolio, then there exist parameters ζ_k such that

$$E[Z_{i,k}] = v_i \zeta_k$$

  holds for all $k \in \{0, 1, \ldots, n\}$ and $i \in \{0, 1, \ldots, n\}$.

It thus follows that development patterns for incremental loss ratios exist for each of the sub-portfolios and also for the total portfolio if and only if there exists some c such that the identity

$$\bar{v}_i / \tilde{v}_i = c$$

holds for all $i \in \{0, 1, \ldots, n\}$.

Example. Depending on the choice of the volume measure, different effects arise from the application of the additive method in *motor third party liability insurance*: If the number of contracts is chosen as the volume measure, the separation of bodily injury claims and pure property damage claims can be omitted, as the same volume measure is used for both types of losses and since the additive method is additive in this case.[1] By contrast, if the corresponding expected number of claims is chosen as the volume measure, then the volume measures for bodily injury claims and for pure property damage claims are usually not proportional and in this case the additive method is not additive in general.

---

[1]Let w_i denote the number of contracts in accident year i. Then one has $\bar{v}_i = w_i$ and $\tilde{v}_i = w_i$, and hence $v_i = 2w_i$. The additive method applied to either v_i and w_i produces the same results since scaling of the volume measures does not affect the predictors.

One possibility of a consistent modelling of the sub-portfolios and the total portfolio is provided by the *multivariate additive model*, which provides a justification of the *multivariate additive method*.

The remarks made on the multivariate chain ladder method also apply to the multivariate additive method.

Remarks

Assertion (1) of both theorems essentially states that, for every future incremental loss, the sum of the predictors from the sub-portfolios is always greater than the predictor from the total portfolio when one of the two sub-portfolios has at the same time a lower *development speed* and a higher *expansion speed* than the other. Similar interpretations can be given for assertions (2) and (3) of these theorems. The expansion over accident years is sometimes called *accident year inflation*.

The theoretical results of this article provide sufficient conditions for underestimation or overestimation of the reserves caused by the aggregation of sub-portfolios. Presumably, in actuarial practice these conditions will only be checked once the predictors have been computed and compared. If, however, it then turns out that the appropriate sufficient condition is fulfilled, then this check provides some useful information on the sub-portfolios.

Notes

Keywords: Additive Method, Chain Ladder Method (Basics), Development Patterns (Basics), Multivariate Methods, Volume Measures.

References: Ajne [1994], Barnett, Zehnwirth & Dubossarski [2005], Fuchs [2014], Klemmt [2005], Schmidt [2006b, 2012].

Bornhuetter–Ferguson Method

Anja Schnaus

Consider the run-off square of cumulative losses:

Accident year	Development year							
	0	1	... k	... $n-i$	... $n-1$	n		
0	$S_{0,0}$	$S_{0,1}$	... $S_{0,k}$	... $S_{0,n-i}$	... $S_{0,n-1}$	$S_{0,n}$		
1	$S_{1,0}$	$S_{1,1}$	... $S_{1,k}$	... $S_{1,n-i}$	... $S_{1,n-1}$	$S_{1,n}$		
$\vdots$	$\vdots$	$\vdots$	$\vdots$	$\vdots$	$\vdots$	$\vdots$		
i	$S_{i,0}$	$S_{i,1}$	... $S_{i,k}$	... $S_{i,n-i}$	... $S_{i,n-1}$	$S_{i,n}$		
$\vdots$	$\vdots$	$\vdots$	$\vdots$	$\vdots$	$\vdots$	$\vdots$		
$n-k$	$S_{n-k,0}$	$S_{n-k,1}$	... $S_{n-k,k}$	... $S_{n-k,n-i}$	... $S_{n-k,n-1}$	$S_{n-k,n}$		
$\vdots$	$\vdots$	$\vdots$	$\vdots$	$\vdots$	$\vdots$	$\vdots$		
$n-1$	$S_{n-1,0}$	$S_{n-1,1}$	... $S_{n-1,k}$	... $S_{n-1,n-i}$	... $S_{n-1,n-1}$	$S_{n-1,n}$		
n	$S_{n,0}$	$S_{n,1}$	... $S_{n,k}$	... $S_{n,n-i}$	... $S_{n,n-1}$	$S_{n,n}$		

We assume that the cumulative losses $S_{i,k}$ are observable for $i + k \leq n$ and that they are non-observable for $i + k \geq n + 1$. For $i, k \in \{0, 1, \ldots, n\}$ we denote by

$$Z_{i,k} := \begin{cases} S_{i,0} & \text{if } k = 0 \\ S_{i,k} - S_{i,k-1} & \text{else} \end{cases}$$

the incremental loss from accident year i in development year k.

A. Schnaus (✉)
Kerpen, Germany
e-mail: schnaus@genre.com

© Springer International Publishing Switzerland 2016
M. Radtke et al. (eds.), *Handbook on Loss Reserving*,
EAA Series, DOI 10.1007/978-3-319-30056-6_3

The Bornhuetter–Ferguson method is a prediction method based on the development pattern for quotas:

Development Pattern for Quotas: *There exist parameters* $\gamma_0, \gamma_1, \ldots, \gamma_n$ *with* $\gamma_n = 1$ *such that the identity*

$$\frac{E[S_{i,k}]}{E[S_{i,n}]} = \gamma_k$$

holds for all $k \in \{0, 1, \ldots, n\}$ *and for all* $i \in \{0, 1, \ldots, n\}$.

In this article we assume that a development pattern for quotas exists. Then the parameters $\vartheta_0, \vartheta_1, \ldots, \vartheta_n$ with

$$\vartheta_k := \begin{cases} \gamma_0 & \text{if } k = 0 \\ \gamma_k - \gamma_{k-1} & \text{else} \end{cases}$$

form a *development pattern for incremental quotas*. In particular, we are in the situation of a *multiplicative model*.

The Bornhuetter–Ferguson method predicts the future cumulative losses $S_{i,k}$ on the basis of the current losses $S_{i,n-i}$ using *a priori estimators*

$$\widehat{\gamma}_0, \widehat{\gamma}_1, \ldots, \widehat{\gamma}_n$$

(*percentage reported*) with $\widehat{\gamma}_n := 1$ of the *quotas* $\gamma_0, \gamma_1, \ldots, \gamma_n$ and *a priori estimators*

$$\widehat{\alpha}_0, \widehat{\alpha}_1, \ldots, \widehat{\alpha}_n$$

(*Bornhuetter–Ferguson run-off losses*) of the expected ultimate losses $\alpha_0, \alpha_1, \ldots, \alpha_n$. The a priori estimators can be based

- exclusively on external information (*benchmarks*), which is not contained in the run-off triangle under consideration (for example on market statistics, on similar portfolios or on pricing assumptions), or
- exclusively on internal information, which is completely contained in the run-off triangle under consideration (for example on weighted means of the observable *individual development factors* $S_{i,k}/S_{i,k-1}$ as estimators of the *development factors* $\varphi_k := \gamma_k/\gamma_{k-1}$), or
- on a combination of external and internal information (for example on weighted means of the scaled cumulative losses $S_{i,k}/\widehat{\gamma}_k$ as estimators of the expected ultimate losses $E[S_{i,n}]$).

The a priori estimators are not subject to any limitation regarding the used information and are assumed as given.

The Bornhuetter–Ferguson method predicts, for every accident year i and every development year k such that $i + k \geq n + 1$, the future cumulative loss $S_{i,k}$ by the *Bornhuetter–Ferguson predictor*

$$S_{i,k}^{\text{BF}} := S_{i,n-i} + \left(\widehat{\gamma}_k - \widehat{\gamma}_{n-i}\right) \widehat{\alpha}_i$$

The definition of the Bornhuetter–Ferguson predictors of cumulative losses replicates
the identity

$$E[S_{i,k}] = E[S_{i,n-i}] + \left(\frac{E[S_{i,k}]}{E[S_{i,n}]} - \frac{E[S_{i,n-i}]}{E[S_{i,n}]} \right) E[S_{i,n}]$$
$$= E[S_{i,n-i}] + (\gamma_k - \gamma_{n-i}) \alpha_i$$

which results from the development pattern for quotas.

From the Bornhuetter–Ferguson predictors of the future cumulative losses one
obtains first the *Bornhuetter–Ferguson predictors*

$$Z_{i,k}^{\mathrm{BF}} := \begin{cases} S_{i,n-i+1}^{\mathrm{BF}} - S_{i,n-i} & \text{if } i + k = n + 1 \\ S_{i,k}^{\mathrm{BF}} - S_{i,k-1}^{\mathrm{BF}} & \text{else} \end{cases}$$

of the future incremental losses $Z_{i,k}$ with $i, k \in \{1, \dots, n\}$ and $i + k \geq n + 1$ and
then the *Bornhuetter–Ferguson predictors*

$$R_i^{\mathrm{BF}} := \sum_{l=n-i+1}^{n} Z_{i,l}^{\mathrm{BF}}$$

$$R_{(c)}^{\mathrm{BF}} := \sum_{l=c-n}^{n} Z_{c-l,l}^{\mathrm{BF}}$$

$$R^{\mathrm{BF}} := \sum_{l=1}^{n} \sum_{j=n-l+1}^{n} Z_{j,l}^{\mathrm{BF}}$$

of the accident year reserves R_i, with $i \in \{1, \dots, n\}$, of the calendar year reserves
$R_{(c)}$, with $c \in \{n+1, \dots, 2n\}$, and of the aggregate loss reserve R. These reserve
predictors are also called *Bornhuetter–Ferguson reserves*.

Because of the identity

$$Z_{i,k}^{\mathrm{BF}} = \left(\widehat{\gamma}_k - \widehat{\gamma}_{k-1} \right) \widehat{\alpha}_i$$

all Bornhuetter–Ferguson reserves are completely determined by the a priori estimators. This identity also provides a short-cut calculation of the Bornhuetter–Ferguson
predictors of the incremental losses.

Example A. Calculation of the Bornhuetter–Ferguson predictors of cumulative losses:

Accident year i	Development year k						$\widehat{\alpha}_i$
	0	1	2	3	4	5	
0						3483	
1					3844	3970	4200
2				3977	4313	4457	4800
3			3880	4690	5068	5230	5400
4		3261	4461	5361	5781	5961	6000
5	1889	3539	4859	5849	6311	6509	6600
$\widehat{\gamma}_k$	0.30	0.55	0.75	0.90	0.97	1	

Incremental losses:

Accident year i	Development year k						$\widehat{\alpha}_i$
	0	1	2	3	4	5	
0							
1						126	4200
2					336	144	4800
3				810	378	162	5400
4			1200	900	420	180	6000
5		1650	1320	990	462	198	6600
$\widehat{\vartheta}_k$	0.30	0.25	0.20	0.15	0.07	0.03	

Reserves:

Accident year i	Reserve R_i^{BF}
1	126
2	480
3	1350
4	2700
5	4620
sum	9276

Calendar year c	Reserve $R_{(c)}^{\mathrm{BF}}$
6	4122
7	2742
8	1572
9	642
10	198
sum	9276

The estimators of the development pattern for incremental quotas are not required for the Bornhuetter–Ferguson method and serve only for comparison with other methods.

Example B. In this example the cumulative loss $S_{4,1}$ is increased by 1000:

Accident	Development year k						$\widehat{\alpha}_i$
year i	0	1	2	3	4	5	
0						3483	
1					3844	3970	4200
2				3977	4313	4457	4800
3			3880	4690	5068	5230	5400
4		4261	5461	6361	6781	6961	6000
5	1889	3539	4859	5849	6311	6509	6600
$\widehat{\gamma}_k$	0.30	0.55	0.75	0.90	0.97	1	

Incremental losses:

Accident	Development year k						$\widehat{\alpha}_i$
year i	0	1	2	3	4	5	
0							
1						126	4200
2					336	144	4800
3				810	378	162	5400
4			1200	900	420	180	6000
5		1650	1320	990	462	198	6600
$\widehat{\vartheta}_k$	0.30	0.25	0.20	0.15	0.07	0.03	

Reserves:

Accident year i	Reserve R_i^{BF}
1	126
2	480
3	1350
4	2700
5	4620
sum	9276

Calander year c	Reserve $R_{(c)}^{\mathrm{BF}}$
6	4122
7	2742
8	1572
9	642
10	198
sum	9276

The predictors of incremental losses and reserves are not changed by the outlier $S_{4,1}$.

Because of the identity

$$S_{i,n}^{\mathrm{BF}} = \widehat{\gamma}_{n-i}\, S_{i,n}^{\mathrm{LD}} + \left(1 - \widehat{\gamma}_{n-i}\right) \widehat{\alpha}_i$$

the Bornhuetter–Ferguson predictors of ultimate losses can be represented as a weighted mean of the *loss development predictors*

$$S_{i,n}^{\mathrm{LD}} := \frac{S_{i,n-i}}{\widehat{\gamma}_{n-i}}$$

and of the a priori estimators $\widehat{\alpha}_i$ of the expected ultimate losses. Thus, in the case $\widehat{\alpha}_i := S_{i,n}^{\mathrm{LD}}$ one obtains $S_{i,n}^{\mathrm{BF}} = S_{i,n}^{\mathrm{LD}}$ and even $S_{i,k}^{\mathrm{BF}} = S_{i,k}^{\mathrm{LD}}$ for all future cumulative losses, which means that in this case the Bornhuetter–Ferguson method coincides with the loss development method.

Because of the manifold possibilities for the choice of a priori estimators of the expected ultimate losses, the Bornhuetter–Ferguson method is more flexible than the loss development method. This freedom of choice has advantages and disadvantages:

- Advantage: With an appropriate choice of the a priori estimators of the expected ultimate losses an overreaction of the Bornhuetter–Ferguson predictors to unusual current losses can be avoided.
- Advantage: The Bornhuetter–Ferguson predictors of incremental losses and reserves are not influenced by the current losses when the a priori estimators are based on external information.
- Disadvantage: The Bornhuetter–Ferguson predictors of the most recent accident years are primarily determined by the a priori estimators of the expected ultimate losses.
- Disadvantage: The use of a priori estimators based on external information implies that the future loss development has no relation at all with the observable cumulative losses.

The disadvantages are mitigated when the a priori estimators also involve information from the run-off triangle.

The Bornhuetter–Ferguson method can be applied independently of the run-off triangle when comparable portfolios or market statistics are used for the a priori estimators of the quotas and of the expected ultimate losses. It is especially appropriate

- for new portfolios, for which no or only very little data is available, and
- for volatile portfolios with a sparsely populated run-off triangle or with a heavily distorted main diagonal or with *large losses*.

An important application area of the Bornhuetter–Ferguson method is *non-proportional reinsurance*.

The representation of the Bornhuetter–Ferguson method given above is an extension of its original version and leads to the *Bornhuetter–Ferguson principle*, which enables a uniform representation of a multitude of prediction methods in loss reserving.

Original Version of the Bornhuetter–Ferguson Method

The original version of the Bornhuetter–Ferguson method only aims at the prediction of the accident year reserves R_i with $i \in \{1, \dots, n\}$ and uses the Bornhuetter–Ferguson predictors

$$R_i^{\mathrm{BF}} = \left(1 - \widehat{\gamma}_{n-i}\right) \widehat{\alpha}_i$$

with a special choice of the a priori estimators:

- The a priori estimators of the quotas γ_{n-i} are given by the *chain ladder quotas*

$$\gamma_{n-i}^{\mathrm{CL}} := \prod_{l=n-i+1}^{n} \frac{1}{\varphi_l^{\mathrm{CL}}}$$

with the *chain ladder factors*

$$\varphi_k^{\mathrm{CL}} := \frac{\sum_{j=0}^{n-k} S_{j,k}}{\sum_{j=0}^{n-k} S_{j,k-1}}$$

- The a priori estimators of the expected cumulative ultimate losses α_i are given by

$$\widehat{\alpha}_i := v_i \widehat{\kappa}$$

with known volume measures v_i of the accident years and an estimator $\widehat{\kappa}$ of the *expected ultimate loss ratios*

$$\kappa_i := E\left[\frac{S_{i,n}}{v_i}\right]$$

which correspond to the volume measures and are assumed to be identical for all accident years. The estimator $\widehat{\kappa}$ is determined *ad hoc* or on the basis of the *individual chain ladder ultimate loss ratios*

$$\kappa_{i,k}^{\mathrm{CL}} := \frac{S_{i,k}}{v_i \gamma_k^{\mathrm{CL}}}$$

with $i + k \leq n$.

Therefore, the original version of the Bornhuetter–Ferguson method is based on the *Cape Cod model* and the use of chain ladder quotas.

Example A. Calculation of the chain ladder quotas:

Accident	Development year k						Volume
year i	0	1	2	3	4	5	v_i
0	1001	1855	2423	2988	3335	3483	4025
1	1113	2103	2774	3422	3844		4456
2	1265	2433	3233	3977			5315
3	1490	2873	3880				5986
4	1725	3261					6939
5	1889						8158
φ_k^{CL}		1.90	1.33	1.23	1.12	1.04	
γ_k^{CL}	0.27	0.52	0.69	0.85	0.96	1	

Calculation of the individual chain ladder ultimate loss ratios:

Accident	Development year k						Volume
year i	0	1	2	3	4	5	v_i
0	0.90	0.88	0.87	0.87	0.87	0.87	4025
1	0.91	0.90	0.90	0.90	0.90		4456
2	0.87	0.88	0.88	0.88			5315
3	0.91	0.92	0.93				5986
4	0.90	0.90					6939
5	0.84						8158

Therewith it seems justifiable to estimate the expected ultimate loss ratios *ad hoc* by the value $\widehat{\kappa} = 0.90$ and to use 90 % of the premiums as a priori estimators of the expected ultimate losses. Using the chain ladder quotas as a priori estimators of the quotas we obtain the Bornhuetter–Ferguson predictors of the future cumulative losses:

Accident	Development year k						Volume
year i	0	1	2	3	4	5	$v_i\widehat{\kappa}$
0						3483	
1					3844	4014	4010
2				3977	4468	4671	4784
3			3880	4748	5300	5529	5387
4		3261	4333	5339	5980	6245	6245
5	1889	3704	4965	6148	6901	7213	7342
γ_k^{CL}	0.27	0.52	0.69	0.85	0.96	1	

Alternatively, the expected ultimate loss ratio of all accident years can be estimated by the *chain ladder ultimate loss ratio*

$$\kappa^{CL} := \frac{\sum_{j=0}^{n} S_{j,n-j}}{\sum_{j=0}^{n} v_j \gamma_{n-j}^{CL}} = \sum_{j=0}^{n} \frac{v_j \gamma_{n-j}^{CL}}{\sum_{h=0}^{n} v_h \gamma_{n-h}^{CL}} \frac{S_{j,n-j}}{v_j \gamma_{n-j}^{CL}} = \sum_{j=0}^{n} \frac{v_j \gamma_{n-j}^{CL}}{\sum_{h=0}^{n} v_h \gamma_{n-h}^{CL}} \kappa_{j,n-j}^{CL}$$

which leads to the Cape Cod method with chain ladder quotas; in this case, however, only the current individual chain ladder ultimate loss ratios $\kappa_{j,n-j}^{CL}$ (on the main diagonal) are used.

Example B. In this example the cumulative loss $S_{4,1}$ is increased by 1000:

Accident year i	Development year k						Volume v_i
	0	1	2	3	4	5	
0	1001	1855	2423	2988	3335	3483	4025
1	1113	2103	2774	3422	3844		4456
2	1265	2433	3233	3977			5315
3	1490	2873	3880				5986
4	1725	4261					6939
5	1889						8158
φ_k^{CL}		2.05	1.33	1.23	1.12	1.04	
γ_k^{CL}	0.25	0.52	0.69	0.85	0.96	1	

The outlier $S_{4,1}$ causes a change of the chain ladder quota γ_0^{CL}.

Accident year i	Development year k						Volume v_i
	0	1	2	3	4	5	
0	0.98	0.88	0.87	0.87	0.87	0.87	4025
1	0.98	0.90	0.90	0.90	0.90		4456
2	0.93	0.88	0.88	0.88			5315
3	0.98	0.92	0.93				5986
4	0.98	1.18					6939
5	0.91						8158

The outlier $S_{4,1}$ causes a change of the individual chain ladder ultimate loss ratios in accident year 4 and development year 0 so that an *ad hoc* estimation of the expected ultimate losses is not advisable. Here the Cape Cod method also offers an alternative, but then other estimators of the quotas should be used instead of the chain ladder quotas.

Benktander–Hovinen Method

The Benktander–Hovinen method, which is also called the *Benktander method*, *Hovinen method* or *Neuhaus method*, differs from the Bornhuetter–Ferguson method by using the Bornhuetter–Ferguson predictors of the ultimate losses instead of the a priori estimators of the expected ultimate losses.

The Benktander Hovinen method predicts, for every accident year i and every development year k such that $i + k \geq n + 1$, the future cumulative loss $S_{i,k}$ by the *Benktander–Hovinen predictor*

$$S_{i,k}^{\mathrm{BH}} := S_{i,n-i} + \left(\widehat{\gamma}_k - \widehat{\gamma}_{n-i}\right) S_{i,n}^{\mathrm{BF}}$$

Therefore, the Benktander–Hovinen method is subject to the *Bornhuetter–Ferguson principle*.

From the Benktander–Hovinen predictors of the future cumulative losses one obtains first the *Benktander–Hovinen predictors*

$$Z_{i,k}^{\mathrm{BH}} := \begin{cases} S_{i,n-i+1}^{\mathrm{BH}} - S_{i,n-i} & \text{if } i + k = n + 1 \\ S_{i,k}^{\mathrm{BH}} - S_{i,k-1}^{\mathrm{BH}} & \text{else} \end{cases}$$

of the future incremental losses $Z_{i,k}$ with $i, k \in \{0, 1, \ldots, n\}$ and $i + k \geq n + 1$ and then the *Benktander–Hovinen predictors*

$$R_i^{\mathrm{BH}} := \sum_{l=n-i+1}^{n} Z_{i,l}^{\mathrm{BH}}$$

$$R_{(c)}^{\mathrm{BH}} := \sum_{l=c-n}^{n} Z_{c-l,l}^{\mathrm{BH}}$$

$$R^{\mathrm{BH}} := \sum_{l=1}^{n} \sum_{j=n-l+1}^{n} Z_{j,l}^{\mathrm{BH}}$$

of the accident year reserves R_i, with $i \in \{1, \ldots, n\}$, of the calendar year reserves $R_{(c)}$, with $c \in \{n+1, \ldots, 2n\}$, and of the aggregate loss reserve R. These reserve predictors are also called *Benktander–Hovinen reserves*.

Example A. Calculation of the Benktander–Hovinen predictors of cumulative losses:

Accident year i	Development year k						$S_{i,5}^{\mathrm{BF}}$
	0	1	2	3	4	5	
0						3483	
1					3844	3963	3970
2				3977	4289	4423	4457
3			3880	4665	5031	5188	5230
4		3261	4453	5347	5765	5943	5961
5	1889	3516	4818	5794	6250	6445	6509
$\widehat{\gamma}_k$	0.30	0.55	0.75	0.90	0.97	1	

Incremental losses:

Accident	Development year k						$S_{i,5}^{\mathrm{BF}}$
year i	0	1	2	3	4	5	
0							
1						119	3970
2					312	134	4457
3				785	366	157	5230
4			1192	894	417	179	5961
5		1627	1302	976	456	195	6509
$\widehat{\vartheta}_k$	0.30	0.25	0.20	0.15	0.07	0.03	

Reserves:

Accident year i	Reserve R_i^{BH}
1	119
2	446
3	1308
4	2682
5	4556
sum	9111

Calendar year c	Reserve $R_{(c)}^{\mathrm{BH}}$
6	4035
7	2696
8	1551
9	634
10	195
sum	9111

The estimators of the development pattern for incremental quotas are not required for the Benktander–Hovinen method and serve only for comparison with other methods.

Example B. In this example the cumulative loss $S_{4,1}$ is increased by 1000:

Accident	Development year k						$S_{i,5}^{\mathrm{BF}}$
year i	0	1	2	3	4	5	
0						3483	
1					3844	3963	3970
2				3977	4289	4423	4457
3			3880	4665	5031	5188	5230
4		4261	5653	6697	7185	7393	6961
5	1889	3516	4818	5794	6250	6445	6509
$\widehat{\gamma}_k$	0.30	0.55	0.75	0.90	0.97	1	

Incremental losses:

Accident	Development year k						$S_{i,5}^{\mathrm{BF}}$
year i	0	1	2	3	4	5	
0							
1						*119*	*3970*
2					*312*	*134*	*4457*
3				*785*	*366*	*157*	*5230*
4			*1392*	*1044*	*487*	*209*	*6961*
5		*1627*	*1302*	*976*	*456*	*195*	*6509*
$\widehat{\vartheta}_k$	0.30	0.25	0.20	0.15	0.07	0.03	

Reserves:

Accident year i	Reserve R_i^{BH}
1	*119*
2	*446*
3	*1308*
4	*3132*
5	*4556*
sum	*9561*

Calendar year c	Reserve $R_{(c)}^{\mathrm{BH}}$
6	*4235*
7	*2846*
8	*1621*
9	*664*
10	*195*
sum	*9561*

The Benktander–Hovinen method aggravates the effect of the outlier $S_{4,1}$ compared to the Bornhuetter–Ferguson method.

Because of the identity

$$S_{i,k}^{\mathrm{BH}} = \widehat{\gamma}_{n-i}\, S_{i,k}^{\mathrm{LD}} + \left(1 - \widehat{\gamma}_{n-i}\right) S_{i,k}^{\mathrm{BF}}$$

the Benktander–Hovinen predictors of cumulative losses can be represented as a weighted mean of the loss development predictors and the Bornhuetter–Ferguson predictors. This is also true for the Benktander–Hovinen predictors of incremental losses and reserves with identical weights in all cases.

In the case $\widehat{\alpha}_i := S_{i,n}^{\mathrm{LD}}$ one has $S_{i,k}^{\mathrm{BF}} = S_{i,k}^{\mathrm{LD}}$ and hence $S_{i,k}^{\mathrm{BH}} = S_{i,k}^{\mathrm{LD}}$, which means that in this case the Benktander–Hovinen method coincides with the loss development method.

The Benktander–Hovinen method can be understood as the first step of an iteration of the Bornhuetter–Ferguson method, which in every step increases the weight of the current losses against that of the a priori estimators of the expected ultimate losses. This idea leads to the iterated Bornhuetter–Ferguson method.

Iterated Bornhuetter–Ferguson Method

In the m-th iteration with $m \in \mathbb{N}_0$, the iterated Bornhuetter–Ferguson method predicts, for every accident year i and every development year k such that $i + k \geq n + 1$, the future cumulative loss $S_{i,k}$ by the *iterated Bornhuetter–Ferguson predictor* of order m

$$\widehat{S}_{i,k}^{(m)} := \begin{cases} S_{i,n-i} + \left(\widehat{\gamma}_k - \widehat{\gamma}_{n-i}\right)\widehat{\alpha}_i & \text{if } m = 0 \\ S_{i,n-i} + \left(\widehat{\gamma}_k - \widehat{\gamma}_{n-i}\right)\widehat{S}_{i,n}^{(m-1)} & \text{else} \end{cases}$$

This yields $\widehat{S}_{i,k}^{(0)} = S_{i,k}^{\mathrm{BF}}$ and $\widehat{S}_{i,k}^{(1)} = S_{i,k}^{\mathrm{BH}}$.

From the iterated Bornhuetter–Ferguson predictors of the future cumulative losses one obtains first the *iterated Bornhuetter–Ferguson predictors*

$$\widehat{Z}_{i,k}^{(m)} := \begin{cases} \widehat{S}_{i,n-i+1}^{(m)} - S_{i,n-i} & \text{if } i + k = n + 1 \\ \widehat{S}_{i,k}^{(m)} - \widehat{S}_{i,k-1}^{(m)} & \text{else} \end{cases}$$

of the future incremental losses $Z_{i,k}$ with $i, k \in \{0, 1, \ldots, n\}$ and $i + k \geq n + 1$ and then the *iterated Bornhuetter–Ferguson predictors*

$$\widehat{R}_i^{(m)} := \sum_{l=n-i+1}^{n} \widehat{Z}_{i,l}^{(m)}$$

$$\widehat{R}_{(c)}^{(m)} := \sum_{l=c-n}^{n} \widehat{Z}_{c-l,l}^{(m)}$$

$$\widehat{R}^{(m)} := \sum_{l=1}^{n} \sum_{j=n-l+1}^{n} \widehat{Z}_{j,l}^{(m)}$$

of the accident year reserves R_i, with $i \in \{1, \ldots, n\}$, of the calendar year reserves $R_{(c)}$, with $c \in \{n+1, \ldots, 2n\}$, and of the aggregate loss reserve R. These reserve predictors are also called *iterated Bornhuetter–Ferguson reserves* of order m.

For the iterated Bornhuetter–Ferguson predictors of the ultimate losses $S_{i,n}$ induction yields the representation

$$\widehat{S}_{i,n}^{(m)} = \left(1 - \left(1-\widehat{\gamma}_{n-i}\right)^m\right) S_{i,n}^{\mathrm{LD}} + \left(1-\widehat{\gamma}_{n-i}\right)^m S_{i,n}^{\mathrm{BF}}$$

as a weighted mean of the loss development predictors and the Bornhuetter–Ferguson predictors. The identity

$$\widehat{S}_{i,k}^{(m)} = \frac{1 - \widehat{\gamma}_k}{1 - \widehat{\gamma}_{n-i}} S_{i,n-i} + \frac{\widehat{\gamma}_k - \widehat{\gamma}_{n-i}}{1 - \widehat{\gamma}_{n-i}} \widehat{S}_{i,n}^{(m)}$$

then yields the representation

$$\widehat{S}_{i,k}^{(m)} = \left(1 - \left(1-\widehat{\gamma}_{n-i}\right)^m\right) S_{i,k}^{LD} + \left(1-\widehat{\gamma}_{n-i}\right)^m S_{i,k}^{BF}$$

of the iterated Bornhuetter–Ferguson predictors of arbitrary future cumulative losses, hence the representation

$$\widehat{Z}_{i,k}^{(m)} = \left(1 - \left(1-\widehat{\gamma}_{n-i}\right)^m\right) Z_{i,k}^{LD} + \left(1-\widehat{\gamma}_{n-i}\right)^m Z_{i,k}^{BF}$$

of the iterated Bornhuetter–Ferguson predictors of arbitrary future incremental losses, and thus, by summation, analogous representations of the iterated Bornhuetter–Ferguson predictors of arbitrary reserves.

The previous identities show that each of the iterated Bornhuetter–Ferguson predictors is a weighted mean of the corresponding loss development and Bornhuetter–Ferguson predictors with weights that are identical in all cases and depend only on the order m.

In the case $\widehat{\gamma}_{n-i} \in (0, 1)$, the weight of the loss development predictors is increasing with respect to the estimator $\widehat{\gamma}_{n-i}$ and the order m. Furthermore, we have the following result:

Theorem. If $\widehat{\gamma}_{n-i} \in (0, 1)$ holds for all $i \in \{0, 1, \ldots, n-1\}$, then the identity

$$\lim_{m \to \infty} \widehat{Z}_{i,k}^{(m)} = Z_{i,k}^{LD}$$

holds for all $i, k \in \{1, \ldots, n\}$ such that $i + k \geq n + 1$ and the corresponding identities hold for the iterated Bornhuetter–Ferguson predictors of future cumulative losses and reserves.

In the case $\widehat{\alpha}_i := S_{i,n-i}/\widehat{\gamma}_{n-i}$ one has $Z_{i,k}^{BF} = Z_{i,k}^{LD}$ and hence $\widehat{Z}_{i,k}^{(m)} = Z_{i,k}^{LD}$ for all $m \in \mathbb{N}_0$.

Example A. Convergence of the predictors of the iterated Bornhuetter–Ferguson method:

Accident year i	Ultimate losses								$S_{i,5}^{LD}$	$\widehat{\alpha}_i$
	$\widehat{S}_{i,5}^{(0)}$	$\widehat{S}_{i,5}^{(1)}$	$\widehat{S}_{i,5}^{(2)}$	$\widehat{S}_{i,5}^{(3)}$	$\widehat{S}_{i,5}^{(4)}$	$\widehat{S}_{i,5}^{(5)}$	...	$\widehat{S}_{i,5}^{(10)}$ ...		
1	3970	3963	3963	3963	3963	3963	...	3963 ...	3963	4200
2	4457	4423	4419	4419	4419	4419	...	4419 ...	4419	4800
3	5230	5188	5177	5174	5174	5173	...	5173 ...	5173	5400
4	5961	5943	5936	5932	5930	5930	...	5929 ...	5929	6000
5	6509	6445	6401	6369	6348	6332	...	6303 ...	6297	6600

Since the a priori estimators of the development pattern for quotas used here are increasing over the development years, the speed of convergence of the predictors is decreasing over the accident years.

Example B. In this example the cumulative loss $S_{4,1}$ is increased by 1000:

Accident year i	Ultimate losses								$S_{i,5}^{\mathrm{LD}}$	$\widehat{\alpha}_i$
	$\widehat{S}_{i,5}^{(0)}$	$\widehat{S}_{i,5}^{(1)}$	$\widehat{S}_{i,5}^{(2)}$	$\widehat{S}_{i,5}^{(3)}$	$\widehat{S}_{i,5}^{(4)}$	$\widehat{S}_{i,5}^{(5)}$	$\dots$	$\widehat{S}_{i,5}^{(10)}$ $\dots$		
1	3970	3963	3963	3963	3963	3963	$\dots$	3963 $\dots$	3963	4200
2	4457	4423	4419	4419	4419	4419	$\dots$	4419 $\dots$	4419	4800
3	5230	5188	5177	5174	5174	5173	$\dots$	5173 $\dots$	5173	5400
4	6961	7393	7588	7676	7715	7733	$\dots$	7747 $\dots$	7747	6000
5	6509	6445	6401	6369	6348	6332	$\dots$	6303 $\dots$	6297	6600

The outlier $S_{4,1}$ affects only the predictors of accident year 4.

For the comparison with other methods it is useful to compute these examples with

- $\widehat{\gamma}_k := \gamma_k^{\mathrm{AD}}$ (*additive quotas*)
- $\widehat{\gamma}_k := \gamma_k^{\mathrm{CL}}$ (*chain ladder quotas*)
- $\widehat{\gamma}_k := \gamma_k^{\mathrm{PA}}$ (*Panning quotas*)

instead of the a priori estimators used here.

In the iterated Bornhuetter–Ferguson method, the a priori estimators of the expected ultimate losses represent the loss expectation prior to the commencement of the development. These a priori estimators are then adapted to the run-off triangle by the iteration.

Remarks

The Bornhuetter–Ferguson method has its origin in a paper by Bornhuetter and Ferguson (1972) and is highly popular, in particular in the United States.

For a comparison of the Bornhuetter–Ferguson method and the chain ladder method from a Bayesian point of view and with regard to credibility methods we refer to England and Verrall (2002).

In order to take better account of the run-off triangle, Benktander (1976) proposed a weighted mean of the loss development and Bornhuetter–Ferguson reserves. Independently of Benktander, Hovinen (1981) and Neuhaus (1992) proposed the same method.

The iterated Bornhuetter–Ferguson method is also the subject of a paper by Mack (2000).

Notes

Keywords: Additive Method, Bornhuetter–Ferguson Principle, Cape Cod Method, Chain Ladder Method (Basics), Development Patterns (Basics), Development Patterns (Estimation), Loss Development Method, Loss Ratios, Multiplicative Models, Panning Method, Reinsurance.

References: Benktander [1976], Bornhuetter & Ferguson [1972], England & Verrall [2002], Hovinen [1981], Ludwig, Schmeißer & Thänert [2009], Mack [2000, 2002], Neuhaus [1992], Pierson [1994], Schmidt [2009, 2012], Schmidt & Zocher [2008].

Bornhuetter–Ferguson Principle

Klaus D. Schmidt and Mathias Zocher

Consider the run-off square of cumulative losses:

Accident year	Development year							
	0	1	... k	... $n-i$		... $n-1$	n
0	$S_{0,0}$	$S_{0,1}$	... $S_{0,k}$	... $S_{0,n-i}$		... $S_{0,n-1}$	$S_{0,n}$
1	$S_{1,0}$	$S_{1,1}$	... $S_{1,k}$	... $S_{1,n-i}$		... $S_{1,n-1}$	$S_{1,n}$
$\vdots$	$\vdots$	$\vdots$	$\vdots$	$\vdots$		$\vdots$	$\vdots$
i	$S_{i,0}$	$S_{i,1}$	... $S_{i,k}$	... $S_{i,n-i}$		... $S_{i,n-1}$	$S_{i,n}$
$\vdots$	$\vdots$	$\vdots$	$\vdots$	$\vdots$		$\vdots$	$\vdots$
$n-k$	$S_{n-k,0}$	$S_{n-k,1}$	... $S_{n-k,k}$	... $S_{n-k,n-i}$	...	$S_{n-k,n-1}$	$S_{n-k,n}$
$\vdots$	$\vdots$	$\vdots$	$\vdots$	$\vdots$		$\vdots$	$\vdots$
$n-1$	$S_{n-1,0}$	$S_{n-1,1}$	... $S_{n-1,k}$	... $S_{n-1,n-i}$	...	$S_{n-1,n-1}$	$S_{n-1,n}$
n	$S_{n,0}$	$S_{n,1}$	... $S_{n,k}$	... $S_{n,n-i}$	...	$S_{n,n-1}$	$S_{n,n}$

We assume that the cumulative losses $S_{i,k}$ are observable for $i + k \leq n$ and that they are non-observable for $i + k \geq n + 1$. For $i, k \in \{0, 1, \ldots, n\}$ we denote by

$$Z_{i,k} := \begin{cases} S_{i,0} & \text{if } k = 0 \\ S_{i,k} - S_{i,k-1} & \text{else} \end{cases}$$

the incremental loss from accident year i in development year k.

K.D. Schmidt (✉) · M. Zocher
Technische Universität Dresden, Dresden, Germany
e-mail: klaus.d.schmidt@tu-dresden.de

© Springer International Publishing Switzerland 2016
M. Radtke et al. (eds.), *Handbook on Loss Reserving*,
EAA Series, DOI 10.1007/978-3-319-30056-6_4

In this article we assume that a development pattern for quotas exists:

Development Pattern for Quotas: *There exist parameters* $\gamma_0, \gamma_1, \ldots, \gamma_n$ *with* $\gamma_n = 1$ *such that the identity*

$$\frac{E[S_{i,k}]}{E[S_{i,n}]} = \gamma_k$$

holds for all $k \in \{0, 1, \ldots, n\}$ *and for all* $i \in \{0, 1, \ldots, n\}$.

Furthermore we consider estimators $\widehat{\gamma}_0, \widehat{\gamma}_1, \ldots, \widehat{\gamma}_n$ with $\widehat{\gamma}_n := 1$ of the *quotas* $\gamma_0, \gamma_1, \ldots, \gamma_n$ and estimators $\widehat{\alpha}_0, \widehat{\alpha}_1, \ldots, \widehat{\alpha}_n$ of the expected ultimate losses $\alpha_0, \alpha_1, \ldots, \alpha_n$. We denote the collections of these estimators by $\widehat{\gamma}$ and $\widehat{\alpha}$, respectively, and we denote by

$$\mathbf{S}^{\mathrm{BF}}(\widehat{\gamma}, \widehat{\alpha})$$

the collection of the *Bornhuetter–Ferguson predictors*

$$S^{\mathrm{BF}}_{i,k}(\widehat{\gamma}, \widehat{\alpha}) := S_{i,n-i} + \left(\widehat{\gamma}_k - \widehat{\gamma}_{n-i}\right) \widehat{\alpha}_i$$

of the future cumulative losses $S_{i,k}$ with $i + k \geq n + 1$. We also use corresponding notation for the predictors of other methods.

The *Bornhuetter–Ferguson principle* consists of an analytical part and a synthetical part:

- The *analytical part* consists in the representation of the predictors of different methods of loss reserving in the form of Bornhuetter–Ferguson predictors with appropriate choices of the estimators $\widehat{\gamma}$ and $\widehat{\alpha}$.
- The *synthetical part* consists in the construction of alternative methods by using novel combinations of estimators in the Bornhuetter–Ferguson predictors.

The comparative application of different methods is quite common in practice. The Bornhuetter–Ferguson principle systemizes this approach, and this systematization provides a framework not only for the comparison of different methods but also for the analysis of the impact of different sources of information used in these methods.

Loss Development Method and Chain Ladder Method

The *loss development predictors*

$$S^{\mathrm{LD}}_{i,k}(\widehat{\gamma}) := \widehat{\gamma}_k \frac{S_{i,n-i}}{\widehat{\gamma}_{n-i}}$$

of the future cumulative losses can be represented in the form

$$S^{\mathrm{LD}}_{i,k}(\widehat{\gamma}) = S_{i,n-i} + \left(\widehat{\gamma}_k - \widehat{\gamma}_{n-i}\right) \alpha^{\mathrm{LD}}_i(\widehat{\gamma})$$

by means of the estimators

$$\alpha_i^{\mathrm{LD}}(\widehat{\gamma}) := \frac{S_{i,n-i}}{\widehat{\gamma}_{n-i}}$$

of the expected ultimate losses. This yields

$$\mathbf{S}^{\mathrm{LD}}(\widehat{\gamma}) = \mathbf{S}^{\mathrm{BF}}(\widehat{\gamma}, \alpha^{\mathrm{LD}}(\widehat{\gamma}))$$

Therefore, the *loss development method* is a special case of the *Bornhuetter–Ferguson method*.

The *chain ladder predictors*

$$S_{i,k}^{\mathrm{CL}} := S_{i,n-i} \prod_{l=n-i+1}^{k} \varphi_l^{\mathrm{CL}}$$

with the *chain ladder factors*

$$\varphi_k^{\mathrm{CL}} := \frac{\sum_{j=0}^{n-k} S_{j,k}}{\sum_{j=0}^{n-k} S_{j,k-1}}$$

can be represented in the form

$$S_{i,k}^{\mathrm{CL}} = \gamma_k^{\mathrm{CL}} \frac{S_{i,n-i}}{\gamma_{n-i}^{\mathrm{CL}}}$$

by means of the *chain ladder quotas*

$$\gamma_k^{\mathrm{CL}} := \prod_{l=k+1}^{n} \frac{1}{\varphi_l^{\mathrm{CL}}}$$

This yields

$$\mathbf{S}^{\mathrm{CL}} = \mathbf{S}^{\mathrm{LD}}\left(\gamma^{\mathrm{CL}}\right) = \mathbf{S}^{\mathrm{BF}}\left(\gamma^{\mathrm{CL}}, \alpha^{\mathrm{LD}}\left(\gamma^{\mathrm{CL}}\right)\right)$$

Therefore, the *chain ladder method* is a special case of the *loss development method*, and hence also of the *Bornhuetter–Ferguson method*.

Cape Cod Method and Additive Method

The Cape Cod method as well as the additive method use known *volume measures* $v_0, v_1, \ldots, v_n$ of the accident years. We denote the collection of these volume measures by $\mathbf{v}$.

The *Cape Cod predictors*

$$S_{i,k}^{CC}(\mathbf{v}, \widehat{\boldsymbol{\gamma}}) := S_{i,n-i} + \left(\widehat{\gamma}_k - \widehat{\gamma}_{n-i}\right) v_i \, \kappa^{CC}(\mathbf{v}, \widehat{\boldsymbol{\gamma}})$$

with the *Cape Cod ultimate loss ratio*

$$\kappa^{CC}(\mathbf{v}, \widehat{\boldsymbol{\gamma}}) := \frac{\sum_{j=0}^n S_{j,n-j}}{\sum_{j=0}^n v_j \widehat{\gamma}_{n-j}}$$

can be represented in the form

$$S_{i,k}^{CC}(\mathbf{v}, \widehat{\boldsymbol{\gamma}}) = S_{i,n-i} + \left(\widehat{\gamma}_k - \widehat{\gamma}_{n-i}\right) \alpha_i^{CC}(\mathbf{v}, \widehat{\boldsymbol{\gamma}})$$

by means of the estimators

$$\alpha_i^{CC}(\mathbf{v}, \widehat{\boldsymbol{\gamma}}) := v_i \, \kappa^{CC}(\mathbf{v}, \widehat{\boldsymbol{\gamma}})$$

of the expected ultimate losses. This yields

$$\mathbf{S}^{CC}(\mathbf{v}, \widehat{\boldsymbol{\gamma}}) = \mathbf{S}^{BF}(\widehat{\boldsymbol{\gamma}}, \boldsymbol{\alpha}^{CC}(\mathbf{v}, \widehat{\boldsymbol{\gamma}}))$$

Therefore, the *Cape Cod method* is a special case of the *Bornhuetter–Ferguson method*.

The *additive predictors*

$$S_{i,k}^{AD}(\mathbf{v}) := S_{i,n-i} + v_i \sum_{l=n-i+1}^{k} \zeta_l^{AD}(\mathbf{v})$$

with the *additive incremental loss ratios*

$$\zeta_k^{AD}(\mathbf{v}) := \frac{\sum_{j=0}^{n-k} Z_{j,k}}{\sum_{j=0}^{n-k} v_j}$$

can be represented in the form

$$S_{i,k}^{AD}(\mathbf{v}) = S_{i,n-i} + \left(\gamma_k^{AD}(\mathbf{v}) - \gamma_{n-i}^{AD}(\mathbf{v})\right) \alpha_i^{AD}(\mathbf{v})$$

by means of the *additive quotas*

$$\gamma_k^{AD}(\mathbf{v}) := \frac{\sum_{l=0}^{k} \zeta_l^{AD}(\mathbf{v})}{\sum_{l=0}^{n} \zeta_l^{AD}(\mathbf{v})}$$

and the estimators

$$\alpha_i^{AD}(\mathbf{v}) := v_i \sum_{l=0}^{n} \zeta_l^{AD}(\mathbf{v})$$

of the expected ultimate losses. This yields

$$\mathbf{S}^{AD}(\mathbf{v}) = \mathbf{S}^{BF}(\boldsymbol{\gamma}^{AD}(\mathbf{v}), \boldsymbol{\alpha}^{AD}(\mathbf{v}))$$

Therefore, the *additive method* is a special case of the *Bornhuetter–Ferguson method*. Because of the identity

$$\sum_{l=0}^{n} \zeta_l^{AD}(\mathbf{v}) = \kappa^{CC}(\mathbf{v}, \boldsymbol{\gamma}^{AD}(\mathbf{v}))$$

one has $\boldsymbol{\alpha}^{AD}(\mathbf{v}) = v_i \, \kappa^{CC}(\mathbf{v}, \boldsymbol{\gamma}^{AD}(\mathbf{v})) = \boldsymbol{\alpha}^{CC}(\mathbf{v}, \boldsymbol{\gamma}^{AD}(\mathbf{v}))$ and hence

$$\mathbf{S}^{AD}(\mathbf{v}) = \mathbf{S}^{BF}(\boldsymbol{\gamma}^{AD}(\mathbf{v}), \boldsymbol{\alpha}^{CC}(\mathbf{v}, \boldsymbol{\gamma}^{AD}(\mathbf{v}))) = \mathbf{S}^{CC}(\mathbf{v}, \boldsymbol{\gamma}^{AD}(\mathbf{v}))$$

Therefore, the *additive method* is also a special case of the *Cape Cod method*.

Panning Method

The *Panning predictors*

$$S_{i,k}^{PA} := S_{i,n-i} + Z_{i,0} \sum_{l=n-i+1}^{k} \xi_l^{PA}$$

with the *Panning factors*

$$\xi_k^{PA} := \frac{\sum_{j=0}^{n-k} Z_{j,0} Z_{j,k}}{\sum_{j=0}^{n-k} Z_{j,0}^2}$$

can be represented in the form

$$S_{i,k}^{PA} = S_{i,n-i} + \left(\gamma_k^{PA} - \gamma_{n-i}^{PA} \right) \alpha_i^{PA}$$

by means of the *Panning quotas*

$$\gamma_k^{PA} := \frac{\sum_{l=0}^{k} \xi_l^{PA}}{\sum_{l=0}^{n} \xi_l^{PA}}$$

and the estimators

$$\alpha_i^{\mathrm{PA}} := Z_{i,0} \sum_{l=0}^{n} \xi_l^{\mathrm{PA}}$$

of the expected ultimate losses. This yields

$$\mathbf{S}^{\mathrm{PA}} = \mathbf{S}^{\mathrm{BF}}(\boldsymbol{\gamma}^{\mathrm{PA}}, \boldsymbol{\alpha}^{\mathrm{PA}})$$

Therefore, the *Panning method* is a special case of the *Bornhuetter–Ferguson method*.

Comparison

The special cases of the Bornhuetter–Ferguson method considered before use different types of information about the portfolio to be examined. We refer to

- the information exclusively based on the data of the run-off triangle as *internal information*,
- the information not based on the data of the run-off triangle as *external information* and
- the information consisting of internal as well as external information as *mixed information*.

In particular, volume measures of the portfolio as well as market statistics and data from comparable portfolios are part of external information.

The chain ladder method and the Panning method exclusively use internal information, while the other three methods considered before use mixed information.

The following table gives an overview of the previously considered special cases of the Bornhuetter–Ferguson method.

Estimators of the expected ultimate losses $\widehat{\gamma}$	Estimators of the quotas		
		γ^{CL}	$\gamma^{\mathrm{AD}}(\mathbf{v}) \quad \gamma^{\mathrm{PA}}$
$\widehat{\alpha}$	Bornhuetter–Ferguson Method (general)		
$\alpha^{\mathrm{LD}}(\widehat{\gamma})$	Loss development Method (general)	Chain ladder Method	
$\alpha^{\mathrm{CC}}(\mathbf{v}, \widehat{\gamma})$	Cape Cod Method (general)		Additive Method
α^{PA}	Panning Method (general)		Panning Method

The table should be understood in the sense that the given estimators of the quotas are also used in the estimators of the expected ultimate losses whenever the latter depend on estimators of quotas. For example,

- the chain ladder method is the Bornhuetter–Ferguson method with $\widehat{\gamma} := \gamma^{\mathrm{CL}}$ and $\widehat{\alpha} := \alpha^{\mathrm{LD}}(\gamma^{\mathrm{CL}})$ and
- the additive method is the Bornhuetter–Ferguson method with $\widehat{\gamma} := \gamma^{\mathrm{AD}}(\mathbf{v})$ and $\widehat{\alpha} := \alpha^{\mathrm{CC}}(\mathbf{v}, \gamma^{\mathrm{AD}}(\mathbf{v}))$.

The table contains vacant spaces corresponding to other combinations of the estimators of γ and α.

Simultaneous Application of Several Methods

The simultaneous application of several methods is widespread in practice. Through the representation of several methods in a standard structure, the Bornhuetter–Ferguson principle allows an analysis of the reasons for the different reserves produced by these methods. In particular, the influence of different estimators of the development pattern and that of different estimators of the expected ultimate losses on the reserves becomes apparent. In addition, the impact of different sources of information (internal or external) on the reserves can be analysed. In this way we can, for the portfolio under consideration, identify applicable and non-applicable methods and sources of information, and we obtain a range of plausible reserves from the applicable methods.

When using the Bornhuetter–Ferguson principle, graphical representations are most helpful. The graphics used in the following examples show the aggregate loss reserve and the reserve for calendar year 6 (the first future calendar year) arising from the 16 methods that result from the above table. In the graphics, the *form* of a symbol describes the estimators of the expected ultimate losses and the *colour* describes the estimators of the development pattern.

Example A. We consider the run-off triangle of incremental losses with known volume measures of the accident years and a priori estimators of the development pattern as well as the corresponding run-off triangle of cumulative losses

Accident year i	Development year k						Volume v_i	α_i^{EX}
	0	1	2	3	4	5		
0	1001	854	568	565	347	148	4025	
1	1113	990	671	648	422		4456	4200
2	1265	1168	800	744			5315	4800
3	1490	1383	1007				5986	5400
4	1725	1536					6939	6000
5	1889						8158	6600
γ_k^{EX}	0.30	0.55	0.75	0.90	0.97	1		
0	1001	1855	2423	2988	3335	3483	4025	
1	1113	2103	2774	3422	3844		4456	4200
2	1265	2433	3233	3977			5315	4800
3	1490	2873	3880				5986	5400
4	1725	3261					6939	6000
5	1889						8158	6600

We thus obtain the following representation of the aggregate loss reserve and of the reserve for calendar year 6:

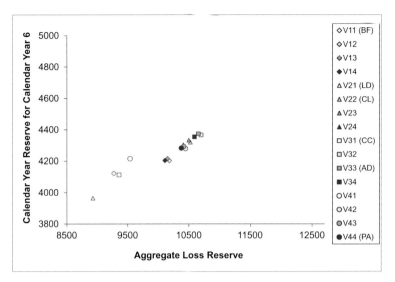

The graphic shows a strong correlation between the aggregate loss reserves and the reserves for calendar year 6. Moreover, all methods using a priori estimators of the development pattern result in the relatively lowest reserves. This indicates that the development in the examined portfolio deviates from the source of the a priori estimators of the development pattern. The other methods lie very closely together, the influence of the estimators of the expected ultimate losses being greater than that of the estimators of the development pattern.

Example B. In this example the incremental loss $Z_{4,1}$ is increased by 1000:

Accident	Development year k						Volume	
year i	0	1	2	3	4	5	v_i	α_i^{EX}
0	1001	854	568	565	347	148	4025	
1	1113	990	671	648	422		4456	4200
2	1265	1168	800	744			5315	4800
3	1490	1383	1007				5986	5400
4	1725	2536					6939	6000
5	1889						8158	6600
γ_k^{EX}	0.30	0.55	0.75	0.90	0.97	1		
0	1001	1855	2423	2988	3335	3483	4025	
1	1113	2103	2774	3422	3844		4456	4200
2	1265	2433	3233	3977			5315	4800
3	1490	2873	3880				5986	5400
4	1725	4261					6939	6000
5	1889						8158	6600

We thus obtain the following representation of the aggregate loss reserve and of the reserve for calendar year 6:

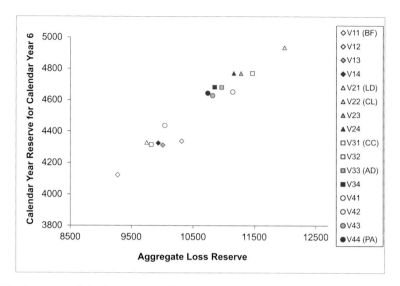

The increase of the incremental loss $Z_{4,1}$ and hence that of the cumulative loss $S_{4,1}$ by 1000 results in a significantly higher volatility of the reserves. Again, all methods using a priori estimators of the development pattern result in the relatively lowest reserves. On the other hand, all methods using chain ladder quotas result in the relatively highest reserves, and also the methods using the loss development estimators of the expected ultimate losses mainly result in relatively high reserves since the outlier on the main diagonal is carried forward in a multiplicative manner. The diagonal sensitivity of the chain ladder method, which provides the highest reserves, is particularly evident and is due to the fact that the outlier affects not only the loss development estimators of the expected ultimate losses but also the chain ladder quotas.

Special importance has to be attached to the case where the volume measures are premiums, because in this case the analysis according to the Bornhuetter–Ferguson principle allows monitoring of the quality of pricing.

Remark

The Bornhuetter–Ferguson principle can be extended to other estimators of the development pattern and of the expected ultimate losses; these can be completely new estimators or simply weighted means of the estimators considered here. Finally, a change of the volume measure leads to changes in all estimators that depend on a volume measure, and one might even use several volume measures simultaneously.

Notes

Keywords: Additive Method, Bornhuetter–Ferguson Method, Cape Cod Method, Chain Ladder Method (Basics), Development Patterns (Basics), Development Patterns (Estimation), Loss Development Method, Multiplicative Models, Panning Method, Volume Measures.

References: GDV [2011], Schmidt [2006a, 2009, 2012], Schmidt & Zocher [2008].

Cape Cod Method

Michael Radtke

Consider the run-off square of cumulative losses:

Accident year	Development year							
	0	1	... k	... $n-i$	... $n-1$	n		
0	$S_{0,0}$	$S_{0,1}$	... $S_{0,k}$	... $S_{0,n-i}$	... $S_{0,n-1}$	$S_{0,n}$		
1	$S_{1,0}$	$S_{1,1}$	... $S_{1,k}$	... $S_{1,n-i}$	... $S_{1,n-1}$	$S_{1,n}$		
$\vdots$	$\vdots$	$\vdots$	$\vdots$	$\vdots$	$\vdots$	$\vdots$		
i	$S_{i,0}$	$S_{i,1}$	... $S_{i,k}$	... $S_{i,n-i}$	... $S_{i,n-1}$	$S_{i,n}$		
$\vdots$	$\vdots$	$\vdots$	$\vdots$	$\vdots$	$\vdots$	$\vdots$		
$n-k$	$S_{n-k,0}$	$S_{n-k,1}$	... $S_{n-k,k}$	... $S_{n-k,n-i}$	... $S_{n-k,n-1}$	$S_{n-k,n}$		
$\vdots$	$\vdots$	$\vdots$	$\vdots$	$\vdots$	$\vdots$	$\vdots$		
$n-1$	$S_{n-1,0}$	$S_{n-1,1}$	... $S_{n-1,k}$	... $S_{n-1,n-i}$	... $S_{n-1,n-1}$	$S_{n-1,n}$		
n	$S_{n,0}$	$S_{n,1}$	... $S_{n,k}$	... $S_{n,n-i}$	... $S_{n,n-1}$	$S_{n,n}$		

We assume that the cumulative losses $S_{i,k}$ are observable for $i + k \leq n$ and that they are non-observable for $i + k \geq n + 1$. For $i, k \in \{0, 1, \ldots, n\}$ we denote by

$$Z_{i,k} := \begin{cases} S_{i,0} & \text{if } k = 0 \\ S_{i,k} - S_{i,k-1} & \text{else} \end{cases}$$

the incremental loss from accident year i in development year k.

The Cape Cod method is a prediction method that uses known *volume measures* $v_0, v_1, \ldots, v_n$ of the accident years and is based on the Cape Cod model:

M. Radtke (✉)
Fachhochschule Dortmund, Dortmund, Germany
e-mail: michael.radtke@fh-dortmund.de

© Springer International Publishing Switzerland 2016
M. Radtke et al. (eds.), *Handbook on Loss Reserving*,
EAA Series, DOI 10.1007/978-3-319-30056-6_5

43

Cape Cod Model: *There exist known volume measures $v_0, v_1, \ldots, v_n$ of the accident years as well as unknown parameters κ and $\gamma_0, \gamma_1, \ldots, \gamma_n$ with $\gamma_n = 1$ such that the identity*

$$E\left[\frac{S_{i,k}}{v_i \gamma_k}\right] = \kappa$$

holds for all $i, k \in \{0, 1, \ldots, n\}$.

In this article we assume that the Cape Cod model is given and we interpret the volume measures occasionally as *premiums*. Because of the identity

$$E[S_{i,k}] = (v_i \kappa)\, \gamma_k$$

we are in the situation of a *multiplicative model*, the parameters $\gamma_0, \gamma_1, \ldots, \gamma_n$ form a *development pattern for quotas*, and the parameters $\vartheta_0, \vartheta_1, \ldots, \vartheta_n$ with

$$\vartheta_k := \begin{cases} \gamma_0 & \text{if } k = 0 \\ \gamma_k - \gamma_{k-1} & \text{else} \end{cases}$$

form a *development pattern for incremental quotas*. Furthermore, the parameter κ is the *expected ultimate loss ratio*, which because of

$$\kappa = E\left[\frac{S_{i,n}}{v_i}\right]$$

is identical for all accident years.

The Cape Cod method predicts the future cumulative losses $S_{i,k}$ on the basis of the current losses $S_{i,n-i}$ using *a priori estimators*

$$\widehat{\gamma}_0, \widehat{\gamma}_1, \ldots, \widehat{\gamma}_n$$

with $\widehat{\gamma}_n := 1$ of the *quotas* $\gamma_0, \gamma_1, \ldots, \gamma_n$. The a priori estimators are not subject to any limitation regarding the used information and are assumed as given.

The Cape Cod method consists of two steps:

• The expected ultimate loss ratio κ is estimated by the *Cape Cod ultimate loss ratio*

$$\kappa^{\mathrm{CC}} := \frac{\sum_{j=0}^n S_{j,n-j}}{\sum_{j=0}^n v_j \widehat{\gamma}_{n-j}}$$

Because of

$$\kappa^{\mathrm{CC}} = \sum_{j=0}^n \frac{v_j \widehat{\gamma}_{n-j}}{\sum_{h=0}^n v_h \widehat{\gamma}_{n-h}} \frac{S_{j,n-j}}{v_j \widehat{\gamma}_{n-j}}$$

the Cape Cod ultimate loss ratio κ^{CC} is a weighted mean of the estimators

$$\widehat{\kappa}_i := \frac{S_{i,n-i}}{v_i \widehat{\gamma}_{n-i}}$$

of the expected ultimate loss ratio κ, with weights being proportional to the *burnt premiums* $v_i \widehat{\gamma}_{n-i}$.

- For every accident year i and every development year k such that $i + k \geq n + 1$, the future cumulative loss $S_{i,k}$ is predicted by the *Cape Cod predictor*

$$S_{i,k}^{CC} := S_{i,n-i} + \left(\widehat{\gamma}_k - \widehat{\gamma}_{n-i}\right) v_i \kappa^{CC}$$

The definition of the Cape Cod predictors of cumulative losses replicates the identity

$$E[S_{i,k}] = E[S_{i,n-i}] + \left(\frac{E[S_{i,k}]}{E[S_{i,n}]} - \frac{E[S_{i,n-i}]}{E[S_{i,n}]}\right) v_i \, E\left[\frac{S_{i,n}}{v_i}\right]$$
$$= E[S_{i,n-i}] + (\gamma_k - \gamma_{n-i}) v_i \kappa$$

which results from the Cape Cod model.

From the Cape Cod predictors of the future cumulative losses one obtains first the *Cape Cod predictors*

$$Z_{i,k}^{CC} := \begin{cases} S_{i,n-i+1}^{CC} - S_{i,n-i} & \text{if } i + k = n + 1 \\ S_{i,k}^{CC} - S_{i,k-1}^{CC} & \text{else} \end{cases}$$

of the future incremental losses $Z_{i,k}$ with $i, k \in \{1, \ldots, n\}$ and $i + k \geq n + 1$ and then the *Cape Cod predictors*

$$R_i^{CC} := \sum_{l=n-i+1}^{n} \widehat{Z}_{i,l}^{CC}$$

$$R_{(c)}^{CC} := \sum_{l=c-n}^{n} \widehat{Z}_{c-l,l}^{CC}$$

$$R^{CC} := \sum_{l=1}^{n} \sum_{j=n-l+1}^{n} \widehat{Z}_{j,l}^{CC}$$

of the accident year reserves R_i, with $i \in \{1, \ldots, n\}$, of the calendar year reserves $R_{(c)}$, with $c \in \{n+1, \ldots, 2n\}$, and of the aggregate loss reserve R. These reserve predictors are also called *Cape Cod reserves*.

Example A. Current losses, volume measures and a priori estimators of the development pattern:

Accident year i	Development year k						Volume v_i
	0	1	2	3	4	5	
0						3483	4025
1					3844		4456
2				3977			5315
3			3880				5986
4		3261					6939
5	1889						8158
$\widehat{\gamma}_k$	0.30	0.55	0.75	0.90	0.97	1	

The volume measures and the a priori estimators of the development pattern yield the burnt premiums:

Accident year i	Development year k						Volume v_i
	0	1	2	3	4	5	
0						4025	4025
1					4322		4456
2				4784			5315
3			4490				5986
4		3816					6939
5	2447						8158
$\widehat{\gamma}_k$	0.30	0.55	0.75	0.90	0.97	1	

The quotient of the sums over the diagonals then yields the Cape Cod ultimate loss ratio $\kappa^{CC} = 0.85$. Cape Cod predictors of cumulative losses:

Accident year i	Development year k						Volume v_i
	0	1	2	3	4	5	
0						3483	4025
1					3844	3958	4456
2				3977	4294	4429	5315
3			3880	4644	5001	5154	5986
4		3261	4443	5329	5742	5919	6939
5	1889	3625	5014	6056	6542	6751	8158
$\widehat{\gamma}_k$	0.30	0.55	0.75	0.90	0.97	1	

Incremental losses:

Accident year i	Development year k						Volume v_i
	0	1	2	3	4	5	
0							4025
1						114	4456
2					317	136	5315
3				764	357	153	5986
4			1182	886	414	177	6939
5		1736	1389	1042	486	208	8158
$\widehat{\vartheta}_k$	0.30	0.25	0.20	0.15	0.07	0.03	

Reserves:

Accident year i	Reserve R_i
1	114
2	452
3	1274
4	2658
5	4862
sum	9361

Calendar year c	Reserve $R_{(c)}$
6	4113
7	2768
8	1608
9	663
10	208
sum	9361

The a priori estimators of the development pattern for incremental quotas are not required for the Cape Cod method and serve only for comparison with other methods.

Example B. In this example the cumulative loss $S_{4,1}$ is increased by 1000:

Accident year i	Development year k						Volume v_i
	0	1	2	3	4	5	
0						3483	4025
1					3844		4456
2				3977			5315
3			3880				5986
4		4261					6939
5	1889						8158
$\widehat{\gamma}_k$	0.30	0.55	0.75	0.90	0.97	1	

The burnt premiums do not change:

Accident year i	Development year k						Volume v_i
	0	1	2	3	4	5	
0						4025	4025
1					4322		4456
2				4784			5315
3			4490				5986
4		3816					6939
5	2447						8158
$\widehat{\gamma}_k$	0.30	0.55	0.75	0.90	0.97	1	

The Cape Cod ultimate loss ratio results now in $\kappa^{CC} = 0.89$. Cape Cod predictors of cumulative losses:

Accident	Development year k						Volume
year i	0	1	2	3	4	5	v_i
0						3483	4025
1					3844	3963	4456
2				3977	4309	4452	5315
3			3880	4682	5056	5217	5986
4		4261	5501	6430	6864	7050	6939
5	1889	3711	5168	6261	6771	6990	8158
$\widehat{\gamma}_k$	0.30	0.55	0.75	0.90	0.97	1	

Incremental losses:

Accident	Development year k						Volume
year i	0	1	2	3	4	5	v_i
0							4025
1						119	4456
2					332	142	5315
3				802	374	160	5986
4			1240	930	434	186	6939
5		1822	1457	1093	510	219	8158
$\widehat{\vartheta}_k$	0.30	0.25	0.20	0.15	0.07	0.03	

Reserves:

Accident year i	Reserve R_i
1	119
2	425
3	1337
4	2789
5	5101
sum	9821

Calendar year c	Reserve $R_{(c)}$
6	4315
7	2904
8	1687
9	696
10	219
sum	9821

Via the Cape Code ultimate loss ratio, the outlier $S_{4,1}$ affects the predictors of all incremental losses and reserves.

For the comparison with other methods it is useful to compute these examples with

- $\widehat{\gamma}_k := \gamma_k^{AD}$ (*additive quotas*)
- $\widehat{\gamma}_k := \gamma_k^{CL}$ (*chain ladder quotas*)
- $\widehat{\gamma}_k := \gamma_k^{PA}$ (*Panning quotas*)

instead of the a priori estimators used here.

Bornhuetter–Ferguson Principle

Letting

$$\alpha_i^{CC} := v_i \kappa^{CC}$$

the Cape Cod predictors of the future cumulative losses can be represented in the form

$$S_{i,k}^{CC} = S_{i,n-i} + \left(\widehat{\gamma}_k - \widehat{\gamma}_{n-i}\right) \alpha_i^{CC}$$

Therefore, the Cape Cod method is subject to the *Bornhuetter–Ferguson principle*.

Furthermore, in the case $\widehat{\gamma}_k := \gamma_k^{AD}$ the Cape Cod ultimate loss ratio satisfies $\kappa^{CC} = \kappa^{AD}$ and the Cape Cod method is equivalent to the *additive method*.

Comparison with the Loss Development Method

Due to its multiplicative basic structure, the *loss development method* causes problems in practice when a cumulative loss on the main diagonal, and hence in the current calendar year, has to be considered as an *outlier* because of certain special effects. Such special effects can arise from large losses or from the absence of losses.

The loss development method is highly diagonal sensitive, due to the fact that in this method the predictors of the future cumulative and incremental losses, and hence those of the reserves, are essentially determined by the current losses. Hence the loss development method is not robust against outliers on the main diagonal.

The diagonal sensitivity of the loss development method is notably strong when the *chain ladder quotas* are used as estimators of the quotas (so that the loss development method becomes the *chain ladder method*). This can be attributed to the fact that the cumulative losses also affect the chain ladder quotas via the *chain ladder factors*.

The Cape Cod method improves the loss development method by cleansing outliers on the main diagonal via a calibration using volume measures of the accident years and the assumption of identical expected ultimate loss ratios over all accident years.

The following consideration illustrates the character of the Cape Cod method and its relation to the loss development method:

According to the definition of the estimators $\widehat{\kappa}_i$ of the expected ultimate loss ratios, the current losses satisfy

$$S_{i,n-i} = \widehat{\gamma}_{n-i} v_i \widehat{\kappa}_i$$

The robustification of these estimators by the Cape Cod ultimate loss ratio κ^{CC} suggests splitting the current losses $S_{i,n-i}$ into the *smoothed current losses*

$$T_{i,n-i} := \widehat{\gamma}_{n-i} v_i \kappa^{CC}$$

and the *outlier effects*

$$X_i := S_{i,n-i} - T_{i,n-i}$$

and then to predict the future smoothed cumulative losses while leaving the outlier effects unchanged. The loss development predictors of the future cumulative losses based on the smoothed current losses satisfy

$$T_{i,k}^{LD} = \widehat{\gamma}_k \frac{T_{i,n-i}}{\widehat{\gamma}_{n-i}} = \widehat{\gamma}_k v_i \kappa^{CC}$$

and for the Cape Cod predictors of the future cumulative losses this yields

$$
\begin{aligned}
S_{i,k}^{CC} &= S_{i,n-i} + \left(\widehat{\gamma}_k - \widehat{\gamma}_{n-i}\right) v_i \kappa^{CC} \\
&= X_i + T_{i,n-i} + \left(\widehat{\gamma}_k - \widehat{\gamma}_{n-i}\right) T_{i,n}^{LD} \\
&= X_i + T_{i,k}^{LD}
\end{aligned}
$$

Therefore, the Cape Cod predictors of the future cumulative losses are the sum of the unchanged outlier effects and the loss development predictors based on the smoothed current losses. Moreover, the outlier effects satisfy

$$\sum_{j=0}^{n} X_j = \sum_{j=0}^{n} S_{j,n-j} - \sum_{j=0}^{n} T_{j,n-j} = 0$$

Furthermore, the Cape Cod predictors of the future incremental losses are identical with the loss development predictors of the future incremental losses based on the smoothed current losses, and consequently the Cape Cod reserves are identical with the loss development reserves based on the smoothed current losses. In conclusion, the Cape Cod method is equivalent to the loss development method after smoothing the current losses by means of the Cape Cod ultimate loss ratio.

Example A. Smoothed current losses, loss development predictors of the future cumulative losses based on the smoothed current losses, and outlier effects:

Accident	Development year k						Outlier effect
year i	0	1	2	3	4	5	X_i
0						3427	56
1					3680	3794	164
2				4072	4389	4525	−95
3			3822	4587	4943	5096	58
4		3249	4431	5317	5730	5908	12
5	2084	3820	5209	6251	6737	6945	−195
$\widehat{\gamma}_k$	0.30	0.55	0.75	0.90	0.97	1	

The sum of the outlier effects is equal to 0.

Example B. In this example the cumulative loss $S_{4,1}$ is increased by 1000:

Accident	Development year k						Outlier effect
year i	0	1	2	3	4	5	X_i
0						3595	−112
1					3861	3980	−17
2				4273	4605	4748	−296
3			4010	4812	5186	5347	−130
4		3409	4649	5578	6012	6198	852
5	2186	4008	5465	6558	7068	7287	−297
$\widehat{\gamma}_k$	0.30	0.55	0.75	0.90	0.97	1	

Because of the outlier $S_{4,1}$ the volatility of the outlier effects increases.

Again, it is also useful to compute these examples for

- $\widehat{\gamma}_k := \gamma_k^{\mathrm{AD}}$ (*additive quotas*)
- $\widehat{\gamma}_k := \gamma_k^{\mathrm{CL}}$ (*chain ladder quotas*)
- $\widehat{\gamma}_k := \gamma_k^{\mathrm{PA}}$ (*Panning quotas*)

instead of the a priori estimators given here.

Remark

The Cape Cod method can be modified by replacing the weights in the Cape Cod ultimate loss ratio

$$\kappa^{\mathrm{CC}} = \sum_{j=0}^{n} \frac{v_j \widehat{\gamma}_{n-j}}{\sum_{h=0}^{n} v_h \widehat{\gamma}_{n-h}} \frac{S_{j,n-j}}{v_j \widehat{\gamma}_{n-j}}$$

by different weights.

Notes

Keywords: Additive Method, Bornhuetter–Ferguson Method, Bornhuetter–Ferguson Principle, Chain Ladder Method (Basics), Development Patterns (Basics), Development Patterns (Estimation), Loss Development Method, Loss Ratios, Multiplicative Models, Volume Measures.

References: Mack [2002], Schmidt [2009, 2012], Schmidt & Zocher [2008], Schweizer Rück [1989], Straub [1988], Taylor [2000].

Chain Ladder Method (Basics)

Klaus D. Schmidt

Consider the run-off square of cumulative losses:

Accident year	Development year						
	0	1	... k	... $n-i$	... $n-1$	n	
0	$S_{0,0}$	$S_{0,1}$	... $S_{0,k}$	... $S_{0,n-i}$	... $S_{0,n-1}$	$S_{0,n}$
1	$S_{1,0}$	$S_{1,1}$	... $S_{1,k}$	... $S_{1,n-i}$	... $S_{1,n-1}$	$S_{1,n}$
$\vdots$	$\vdots$	$\vdots$	$\vdots$	$\vdots$	$\vdots$	$\vdots$
i	$S_{i,0}$	$S_{i,1}$	... $S_{i,k}$	... $S_{i,n-i}$	... $S_{i,n-1}$	$S_{i,n}$
$\vdots$	$\vdots$	$\vdots$	$\vdots$	$\vdots$	$\vdots$	$\vdots$
$n-k$	$S_{n-k,0}$	$S_{n-k,1}$	... $S_{n-k,k}$	... $S_{n-k,n-i}$	... $S_{n-k,n-1}$	$S_{n-k,n}$
$\vdots$	$\vdots$	$\vdots$	$\vdots$	$\vdots$	$\vdots$	$\vdots$
$n-1$	$S_{n-1,0}$	$S_{n-1,1}$	... $S_{n-1,k}$	... $S_{n-1,n-i}$	... $S_{n-1,n-1}$	$S_{n-1,n}$
n	$S_{n,0}$	$S_{n,1}$	... $S_{n,k}$	... $S_{n,n-i}$	... $S_{n,n-1}$	$S_{n,n}$

We assume that the cumulative losses $S_{i,k}$ are observable for $i + k \leq n$ and that they are non-observable for $i + k \geq n + 1$. For $i, k \in \{0, 1, \ldots, n\}$ we denote by

$$Z_{i,k} := \begin{cases} S_{i,0} & \text{if } k = 0 \\ S_{i,k} - S_{i,k-1} & \text{else} \end{cases}$$

the incremental loss from accident year i in development year k.

K.D. Schmidt (✉)
Technische Universität Dresden, Dresden, Germany
e-mail: klaus.d.schmidt@tu-dresden.de

© Springer International Publishing Switzerland 2016
M. Radtke et al. (eds.), *Handbook on Loss Reserving*,
EAA Series, DOI 10.1007/978-3-319-30056-6_6

53

The chain ladder method is a prediction method based on the development pattern for factors:

Development Pattern for Factors: *There exist parameters $\varphi_1, \ldots, \varphi_n$ such that the identity*

$$\frac{E[S_{i,k}]}{E[S_{i,k-1}]} = \varphi_k$$

holds for all $k \in \{1, \ldots, n\}$ and for all $i \in \{0, 1, \ldots, n\}$.

In this article, we assume that there exists a development pattern for factors. Then the parameters $\gamma_0, \gamma_1, \ldots, \gamma_n$ given by

$$\gamma_k := \prod_{l=k+1}^{n} \frac{1}{\varphi_l}$$

form a *development pattern for quotas* and the parameters $\vartheta_0, \vartheta_1, \ldots, \vartheta_n$ given by

$$\vartheta_k := \begin{cases} \gamma_0 & \text{if } k = 0 \\ \gamma_k - \gamma_{k-1} & \text{else} \end{cases}$$

form a *development pattern for incremental quotas*. In particular, we are in the situation of a *multiplicative model*.

The *chain ladder method* consists of two steps:

- For every development year $k \in \{1, \ldots, n\}$, the development factor φ_k is estimated by the *chain ladder factor*

$$\varphi_k^{\mathrm{CL}} := \frac{\sum_{j=0}^{n-k} S_{j,k}}{\sum_{j=0}^{n-k} S_{j,k-1}}$$

Since

$$\varphi_k^{\mathrm{CL}} = \sum_{j=0}^{n-k} \frac{S_{j,k-1}}{\sum_{h=0}^{n-k} S_{h,k-1}} \frac{S_{j,k}}{S_{j,k-1}}$$

the chain ladder factor φ_k^{CL} is a weighted mean of the observable *individual development factors* $S_{i,k}/S_{i,k-1}$ of development year k, with weights proportional to the cumulative losses of the preceding development year and the same accident year.
- For every accident year i and every development year k such that $i + k \geq n + 1$, the future cumulative loss $S_{i,k}$ is predicted by the *chain ladder predictor*

$$S_{i,k}^{\mathrm{CL}} := S_{i,n-i} \prod_{l=n-i+1}^{k} \varphi_l^{\mathrm{CL}}$$

The definition of the chain ladder predictors of the cumulative losses reflects the identity

$$E[S_{i,k}] = E[S_{i,n-i}] \prod_{l=n-i+1}^{k} \frac{E[S_{i,l}]}{E[S_{i,l-1}]} = E[S_{i,n-i}] \prod_{l=n-i+1}^{k} \varphi_l$$

which results from the development pattern for factors.

Using the chain ladder predictors of the future cumulative losses, we define the *chain ladder predictors*

$$Z_{i,k}^{\mathrm{CL}} := \begin{cases} S_{i,n-i+1}^{\mathrm{CL}} - S_{i,n-i} & \text{if } i+k = n+1 \\ S_{i,k}^{\mathrm{CL}} - S_{i,k-1}^{\mathrm{CL}} & \text{else} \end{cases}$$

of the future incremental losses $Z_{i,k}$ and the *chain ladder predictors*

$$R_i^{\mathrm{CL}} := \sum_{l=n-i+1}^{n} Z_{i,l}^{\mathrm{CL}}$$

$$R_{(c)}^{\mathrm{CL}} := \sum_{l=c-n}^{n} Z_{c-l,l}^{\mathrm{CL}}$$

$$R^{\mathrm{CL}} := \sum_{l=1}^{n} \sum_{j=n-l+1}^{n} Z_{j,l}^{\mathrm{CL}}$$

of the accident year reserves R_i, with $i \in \{1, \ldots, n\}$, of calendar year reserves $R_{(c)}$, with $c \in \{n+1, \ldots, 2n\}$, and of the aggregate loss reserve R. These reserve predictors are also called *chain ladder reserves*. Obviously, all chain ladder reserves are completely determined by the run-off triangle.

Example A. Calculation of the chain ladder predictors of cumulative losses:

Accident year i	Development year k					
	0	1	2	3	4	5
0	1001	1855	2423	2988	3335	3483
1	1113	2103	2774	3422	3844	4015
2	1265	2433	3233	3977	4454	4652
3	1490	2873	3880	4781	5354	5592
4	1725	3261	4333	5339	5980	6245
5	1889	3588	4768	5875	6579	6871
φ_k^{CL}		1.90	1.33	1.23	1.12	1.04
γ_k^{CL}	0.27	0.52	0.69	0.85	0.96	1

Incremental losses:

Accident year i	Development year k						Sum
	0	1	2	3	4	5	
0	1001	854	568	565	347	148	3483
1	1113	990	671	648	422	171	4015
2	1265	1168	800	744	477	198	4652
3	1490	1383	1007	901	574	238	5592
4	1725	1536	1072	1006	641	265	6245
5	1889	1699	1180	1107	705	292	6871
$\vartheta_k^{\mathrm{CL}}$	0.27	0.25	0.17	0.16	0.10	0.04	1

Reserves:

Accident year i	Reserve R_i^{CL}
1	171
2	675
3	1712
4	2984
5	4982
sum	10524

Calendar year c	Reserve $R_{(c)}^{\mathrm{CL}}$
6	4320
7	2957
8	1985
9	970
10	292
sum	10524

The estimators of the development patterns for quotas and incremental quotas are not needed in the chain ladder method but serve for comparison with other methods.

Example B. In this example the cumulative loss $S_{4,1}$ is increased by 1000:

Accident year i	Development year k					
	0	1	2	3	4	5
0	1001	1855	2423	2988	3335	3483
1	1113	2103	2774	3422	3844	4015
2	1265	2433	3233	3977	4454	4652
3	1490	2873	3880	4781	5354	5592
4	1725	4261	5662	6976	7813	8160
5	1889	3875	5148	6344	7105	7420
φ_k^{CL}		2.05	1.33	1.23	1.12	1.04
γ_k^{CL}	0.25	0.52	0.69	0.85	0.96	1

Incremental losses:

Accident year i	Development year k						Sum
	0	1	2	3	4	5	
0	1001	854	568	565	347	148	3483
1	1113	990	671	648	422	171	4015
2	1265	1168	800	744	477	198	4652
3	1490	1383	1007	901	574	238	5592
4	1725	2536	1401	1314	837	347	8160
5	1889	1986	1274	1195	761	315	7420
ϑ_k^{CL}	0.25	0.27	0.17	0.16	0.10	0.04	1

Reserves:

Accident year i	Reserve R_i^{CL}
1	171
2	675
3	1712
4	3899
5	5531
sum	11987

Calendar year c	Reserve $R_{(c)}^{CL}$
6	4935
7	3360
8	2270
9	1108
10	315
sum	11987

The outlier $Z_{4,1}$ affects the predictors of all incremental or cumulative losses and of the reserve of accident year 4, and it also affects the estimator of the parameter φ_1 and hence the predictors of all incremental or cumulative losses and of the reserve of accident year 5 and the predictors of all calender year reserves.

Bornhuetter–Ferguson Principle

Using the *chain ladder quotas*

$$\gamma_k^{CL} := \prod_{l=k+1}^{n} \frac{1}{\varphi_l^{CL}}$$

the chain ladder predictors of the future cumulative losses can be represented in the form

$$S_{i,k}^{CL} = \gamma_k^{CL} \frac{S_{i,n-i}}{\gamma_{n-i}^{CL}}$$

Therefore, the chain ladder method is a special case of the *loss development method*. Moreover, using the *chain ladder ultimate losses*

$$\alpha_i^{CL} := \frac{S_{i,n-i}}{\gamma_{n-i}^{CL}}$$

we obtain the representation

$$S_{i,k}^{CL} = S_{i,n-i} + \left(\gamma_k^{CL} - \gamma_{n-i}^{CL}\right)\alpha_i^{CL}$$

Therefore, the chain ladder method is subject to the *Bornhuetter–Ferguson principle*.

Remarks

The chain ladder method provides the same results as the *grossing up method* and the *marginal sum method*.

The structure of the chain ladder method is very similar to that of the *additive method* and of the *Panning method*. Correspondingly, there exists a strong similarity between the chain ladder models of Mack and Schnaus and the *additive model* as well as the *Panning model*.

The chain ladder method can be modified by changing the weights in the chain ladder factors

$$\varphi_k^{CL} = \sum_{j=0}^{n-k} \frac{S_{j,k-1}}{\sum_{h=0}^{n-k} S_{h,k-1}} \frac{S_{j,k}}{S_{j,k-1}}$$

The large variety of possible weights of the observable *individual development factors*

$$\widehat{\varphi}_{i,k} := S_{i,k}/S_{i,k-1}$$

leads to the question of which weights are optimal in a certain sense. Moreover, with regard to the chain ladder factors, one may ask in which sense and under which conditions the *chain ladder weights*

$$\frac{S_{i,k-1}}{\sum_{h=0}^{n-k} S_{h,k-1}}$$

are optimal. This question is examined in the chain ladder models of Mack and Schnaus.

The chain ladder method was probably first described by Tarbell (1934).

Notes

Keywords: Additive Method, Aggregation, Bornhuetter–Ferguson Method, Born-huetter–Ferguson Principle, Chain Ladder Method (Models), Chain Ladder Method (Prediction Error), Development Patterns (Basics), Development Patterns (Estimation), Grossing Up Method, Loss Development Method, Marginal Sum Method, Multiplicative Models, Multivariate Methods, Munich Chain Ladder Method, Panning Method, Poisson Model, Run-Off Triangles.

References: Institute of Actuaries [1989], Lorenz & Schmidt [1999], Mack [1993, 1994a, 1994b, 2002], Mack & Venter [2000], Reich & Zeller [1988], Schmidt [1997, 1999a, 1999b, 2009, 2012], Schmidt & Schnaus [1996], Schmidt & Wünsche [1998], Schmidt & Zocher [2008], Tarbell [1934].

Chain Ladder Method (Models)

Klaus Th. Hess, Klaus D. Schmidt and Anja Schnaus

Consider the run-off square of cumulative losses:

Accident year	Development year						
	0	1	... k	... $n-i$	... $n-1$	n	
0	$S_{0,0}$	$S_{0,1}$	... $S_{0,k}$	... $S_{0,n-i}$	... $S_{0,n-1}$	$S_{0,n}$	
1	$S_{1,0}$	$S_{1,1}$	... $S_{1,k}$	... $S_{1,n-i}$	... $S_{1,n-1}$	$S_{1,n}$	
$\vdots$	$\vdots$	$\vdots$	$\vdots$	$\vdots$	$\vdots$	$\vdots$	
i	$S_{i,0}$	$S_{i,1}$	... $S_{i,k}$	... $S_{i,n-i}$	... $S_{i,n-1}$	$S_{i,n}$	
$\vdots$	$\vdots$	$\vdots$	$\vdots$	$\vdots$	$\vdots$	$\vdots$	
$n-k$	$S_{n-k,0}$	$S_{n-k,1}$	... $S_{n-k,k}$	... $S_{n-k,n-i}$	... $S_{n-k,n-1}$	$S_{n-k,n}$	
$\vdots$	$\vdots$	$\vdots$	$\vdots$	$\vdots$	$\vdots$	$\vdots$	
$n-1$	$S_{n-1,0}$	$S_{n-1,1}$	... $S_{n-1,k}$	... $S_{n-1,n-i}$	... $S_{n-1,n-1}$	$S_{n-1,n}$	
n	$S_{n,0}$	$S_{n,1}$	... $S_{n,k}$	... $S_{n,n-i}$	... $S_{n,n-1}$	$S_{n,n}$	

We assume that the cumulative losses $S_{i,k}$ are observable for $i + k \leq n$ and that they are non-observable for $i + k \geq n + 1$.

In this article we present three sequential models in which assumptions are made on the first and second conditional moments of the cumulative losses, given the

K.Th. Hess
Universität Rostock, Rostock, Germany
e-mail: klaus-thomas.hess@uni-rostock.de

K.D. Schmidt
Technische Universität Dresden, Dresden, Germany
e-mail: klaus.d.schmidt@tu-dresden.de

A. Schnaus (✉)
Kerpen, Germany
e-mail: schnaus@genre.com

© Springer International Publishing Switzerland 2016
M. Radtke et al. (eds.), *Handbook on Loss Reserving*,
EAA Series, DOI 10.1007/978-3-319-30056-6_7

cumulative losses of older development years, and which justify the chain ladder method to a certain extent with regard to unbiasedness or optimality of the chain ladder predictors.

Three types of σ-algebras on the underlying probability space are required in order to formulate the assumptions made in these models:

- For $k \in \{0, 1, \ldots, n\}$ and $i \in \{0, 1, \ldots, n\}$, we denote by

$$\mathcal{F}_{i,k}$$

the σ-algebra generated by the family $\{S_{i,l}\}_{l \in \{0,1,\ldots,k\}}$. This σ-algebra contains the information from accident year i that is available at the end of development year k.
- For $k \in \{0, 1, \ldots, n\}$, we denote by

$$\mathcal{F}_k$$

the σ-algebra generated by the family $\{S_{j,l}\}_{l \in \{0,1,\ldots,k\}, \, j \in \{0,1,\ldots,n\}}$. This σ-algebra contains the information from all accident years that is available at the end of development year k.
- For $k \in \{0, 1, \ldots, n\}$, we denote by

$$\mathcal{G}_k$$

the σ-algebra generated by the family $\{S_{j,l}\}_{l \in \{0,1,\ldots,k\}, \, j \in \{0,1,\ldots,n-k\}}$. This σ-algebra only contains the information from the accident years $j \in \{0, 1, \ldots, n-k\}$ that is available at the end of development year k.

These σ-algebras are related by $\mathcal{F}_{i,k} \subseteq \mathcal{F}_k$ and $\mathcal{G}_k \subseteq \mathcal{F}_k$.

The Chain Ladder Method and Its Variations

The *chain ladder method* predicts the future cumulative losses $S_{i,k}$ by the *chain ladder predictors*

$$S_{i,k}^{CL} := S_{i,n-i} \prod_{l=n-i+1}^{k} \varphi_l^{CL}$$

with the *chain ladder factors*

$$\varphi_k^{CL} := \frac{\sum_{j=0}^{n-k} S_{j,k}}{\sum_{j=0}^{n-k} S_{j,k-1}}$$

The chain ladder factor φ_k^{CL} is determined by the information contained in the σ-algebra $\mathcal{F}_k$ and can be represented in the form

$$\varphi_k^{\mathrm{CL}} = \sum_{j=0}^{n-k} \frac{S_{j,k-1}}{\sum_{h=0}^{n-k} S_{h,k-1}} \, \widehat{\varphi}_{j,k}$$

as a weighted mean of the *individual development factors*

$$\widehat{\varphi}_{i,k} := \frac{S_{i,k}}{S_{i,k-1}}$$

The individual development factor $\widehat{\varphi}_{i,k}$ is determined by the information contained in the σ-algebra $\mathcal{F}_{i,k}$ and the *chain ladder weights*

$$W_{i,k}^{\mathrm{CL}} := \frac{S_{i,k-1}}{\sum_{h=0}^{n-k} S_{h,k-1}}$$

are determined by the information contained in the σ-algebra $\mathcal{F}_{k-1}$.

If, for every $k \in \{1, \ldots, n\}$, the family $\{W_{j,k}^{\mathrm{CL}}\}_{j\in\{0,1,\ldots,n-k\}}$ of chain ladder weights is replaced by another family $\{W_{j,k}\}_{j\in\{0,1,\ldots,n-k\}}$ of random variables that are also determined by the information contained in the σ-algebra $\mathcal{F}_{k-1}$ and satisfy $\sum_{j=0}^{n-k} W_{j,k} = 1$, then one obtains the *modified chain ladder factors*

$$\widehat{\varphi}_k := \sum_{j=0}^{n-k} W_{j,k} \, \widehat{\varphi}_{j,k}$$

such that $\widehat{\varphi}_k$ is determined by the information contained in the σ-algebra $\mathcal{F}_k$, and hence the *modified chain ladder predictors*

$$\widehat{S}_{i,k} := S_{i,n-i} \prod_{l=n-i+1}^{k} \widehat{\varphi}_l$$

of the future cumulative losses $S_{i,k}$.

We now examine, under the assumptions of certain models and for a given future cumulative loss, unbiasedness of the modified chain ladder predictors and optimality of the chain ladder predictors within the class of all modified chain ladder predictors.

Chain Ladder Model of Mack

The chain ladder model of Mack consists of the following assumptions:

Chain Ladder Model of Mack:

  (i) *The accident years are independent.*

  (ii) *There exist unknown parameters $\varphi_1, \ldots, \varphi_n$ and parameters $\sigma_1^2, \ldots, \sigma_n^2$ such that the identities*

$$E(S_{i,k} \mid \mathcal{F}_{i,k-1}) = S_{i,k-1}\,\varphi_k$$
$$\mathrm{var}(S_{i,k} \mid \mathcal{F}_{i,k-1}) = S_{i,k-1}\,\sigma_k^2$$

hold for every $k \in \{1, \ldots, n\}$ and for all $i \in \{0, 1, \ldots, n\}$.

The assumptions of the chain ladder model of Mack imply that

$$E[S_{i,k}] = E[S_{i,k-1}]\,\varphi_k$$

Therefore, the parameters $\varphi_1, \ldots, \varphi_n$ form a *development pattern for factors*, and it follows that there also exists a *development pattern for incremental quotas* and that we are in the situation of a *multiplicative model*.

To estimate the expected ultimate losses in the chain ladder model of Mack, one may thus use the *marginal sum method*. As the marginal sum estimators of the expected ultimate losses are identical with the chain ladder predictors of the ultimate losses, this provides a first connection between the chain ladder model of Mack and the chain ladder method.

Chain Ladder Model of Schnaus

The chain ladder model of Schnaus consists of the following assumptions:

Chain Ladder Model of Schnaus: *There exist unknown parameters $\varphi_1, \ldots, \varphi_n$ and parameters $\sigma_1^2, \ldots, \sigma_n^2$ such that the identities*

$$E(S_{i,k} \mid \mathcal{F}_{k-1}) = S_{i,k-1}\,\varphi_k$$
$$\mathrm{cov}(S_{i,k}, S_{j,k} \mid \mathcal{F}_{k-1}) = S_{i,k-1}\,\sigma_k^2\,\delta_{i,j}$$

hold for every $k \in \{1, \ldots, n\}$ and for all $i, j \in \{0, 1, \ldots, n\}$.

We first establish a relation between the chain ladder models of Mack and Schnaus:

Theorem. *The chain ladder model of Mack fulfills the assumptions of the chain ladder model of Schnaus.*

In the chain ladder model of Schnaus, and hence in that of Mack, one has

$$E(S_{i,k} \mid \mathcal{F}_{k-1}) = S_{i,k-1}\,\varphi_k$$

which can also be written as

$$E\left(\frac{S_{i,k}}{S_{i,k-1}} \,\middle|\, \mathcal{F}_{k-1}\right) = \varphi_k$$

These identities present a refinement of the notion of a development pattern for factors. Indeed, taking expectations yields

$$\frac{E[S_{i,k}]}{E[S_{i,k-1}]} = \varphi_k = E\left[\frac{S_{i,k}}{S_{i,k-1}}\right]$$

Because of the left identity, the parameters $\varphi_1, \ldots, \varphi_n$ form a *development pattern for factors*. Moreover, because of the right identity, the expected individual development factors $E[\widehat{\varphi}_{i,k}] = E[S_{i,k}/S_{i,k-1}]$ are also independent of the accident year i. This shows that in the chain ladder models of Mack and Schnaus the basic idea of a development pattern is formalized more precisely than it is in other models of loss reserving.

The previous identities also yield

$$\mathrm{cov}\left[\frac{S_{i,k}}{S_{i,k-1}}, S_{i,k-1}\right] = 0$$

which means that the individual development factor $\widehat{\varphi}_{i,k} = S_{i,k}/S_{i,k-1}$ and the cumulative loss $S_{i,k-1}$ are uncorrelated in the chain ladder models of Mack and Schnaus.

Unbiased Predictors

In this section we examine unbiasedness of the chain ladder predictors in the chain ladder model of Schnaus.

The starting point of our considerations is the observation that, as noted before, the first identity in the chain ladder model of Schnaus can be written as

$$E(\widehat{\varphi}_{i,k}\,|\,\mathcal{F}_{k-1}) = E\left(\frac{S_{i,k}}{S_{i,k-1}}\,\bigg|\,\mathcal{F}_{k-1}\right) = \varphi_k$$

This is a condition on the individual development factors. For the chain ladder factors this yields

$$E(\varphi_k^{\mathrm{CL}}\,|\,\mathcal{F}_{k-1}) = \varphi_k$$

Using this identity and the tower property of conditional expectations, one obtains

$$E(S_{i,k}\,|\,\mathcal{F}_{n-i}) = E\left(S_{i,n-i}\prod_{l=n-i+1}^{k}\widehat{\varphi}_{i,l}\,\bigg|\,\mathcal{F}_{n-i}\right) = S_{i,n-i}\prod_{l=n-i+1}^{k}\varphi_l$$

for the future cumulative losses and

$$E(S_{i,k}^{\mathrm{CL}}\,|\,\mathcal{F}_{n-i}) = E\left(S_{i,n-i}\prod_{l=n-i+1}^{k}\varphi_l^{\mathrm{CL}}\,\bigg|\,\mathcal{F}_{n-i}\right) = S_{i,n-i}\prod_{l=n-i+1}^{k}\varphi_l$$

for their chain ladder predictors. Taking expectations then yields the following theorem:

Theorem. *In the chain ladder model of Schnaus, the chain ladder predictors are unbiased.*

This result is remarkable since unbiasedness of the chain ladder predictors is an open question in other stochastic models leading to the chain ladder method. However, unbiasedness is not a property that distinguishes the chain ladder predictors from all other predictors:

Theorem. *In the chain ladder model of Schnaus, all modified chain ladder predictors are unbiased.*

This raises the question of whether there is another quality that distinguishes the chain ladder predictors of a given future cumulative loss at least from all other modified chain ladder predictors of the same cumulative loss.

Optimal Predictors

Under the assumptions of the chain ladder models of Mack and Schnaus, we now examine optimality of the chain ladder predictors within the class of all modified chain ladder predictors of a given future cumulative loss. For $k \in \{1, \ldots, n\}$, we denote by

$$\Phi_k$$

the collection of all modified chain ladder factors for development year k and, for $k \in \{1, \ldots, n\}$ and $i \in \{1, \ldots, n - k\}$, we denote by

$$\Delta_{i,k}$$

the collection of all modified chain ladder predictors of the cumulative loss $S_{i,k}$ from accident year i in development year k. Then, for every $\widehat{\varphi}_k \in \Phi_k$, there exists an $\mathcal{F}_{k-1}$-measurable family $\{W_{j,k}\}_{j \in \{0,1,\ldots,n-k\}}$ of random variables such that $\sum_{j=0}^{n-k} W_{j,k} = 1$ and

$$\widehat{\varphi}_k = \sum_{j=0}^{n-k} W_{j,k}\, \widehat{\varphi}_{j,k}$$

and, for every $\widehat{S}_{i,k} \in \Delta_{i,k}$, there exists a family $\{\widehat{\varphi}_l\}_{l \in \{n-i+1,\ldots,k\}}$ of modified chain ladder factors with $\widehat{\varphi}_l \in \Phi_l$ for all $l \in \{n-i+1, \ldots, k\}$ and such that

$$\widehat{S}_{i,k} = S_{i,n-i} \prod_{l=n-i+1}^{k} \widehat{\varphi}_l$$

Recall that $\varphi_k^{CL} \in \Phi_k$ and $S_{i,k}^{CL} \in \Delta_{i,k}$.

The random variables in $\Delta_{i,k}$ are referred to as *admissible predictors* of the cumulative loss $S_{i,k}$. An admissible predictor is called an *optimal predictor* of $S_{i,k}$ if it minimizes the conditional squared prediction error

$$E\left(\left(\widehat{S}_{i,k} - S_{i,k} \right)^2 \, \Big| \, \mathcal{F}_{n-i} \right)$$

over all admissible predictors $\widehat{S}_{i,k} \in \Delta_{i,k}$.

Theorem. *In the chain ladder model of Schnaus, the chain ladder predictor $S_{i,n-i+1}^{CL}$ is an optimal predictor of $S_{i,n-i+1}$.*

This result distinguishes the chain ladder predictors of the cumulative losses of the first non-observable calendar year from all modified chain ladder predictors. However, a corresponding result for later calendar years does not hold:

Theorem. *In the chain ladder model of Mack, the chain ladder predictor $S_{i,n-i+2}^{CL}$ is not necessarily an optimal predictor of $S_{i,n-i+2}$.*

This negative result loses its severity if one takes into account that prediction of cumulative losses in the near future is far more important than prediction of cumulative losses in the more distant future: Predictors for the distant future can be corrected later, and they even should be corrected since the amount of information that can be used for prediction is increasing over time.

These considerations lead to a *sequential optimality criterion* that consists in minimization of the conditional squared prediction error over

- all $\widehat{S}_{i,k} \in \Delta_{i,k}$ (as before), if $k = n-i+1$, and
- all $\widehat{S}_{i,k} \in \Delta_{i,k}$ satisfying

$$\widehat{S}_{i,k} = S_{i,k-1}^{CL} \, \widehat{\varphi}_k$$

for some $\widehat{\varphi}_k \in \Phi_k$, if $k \in \{n-i+2, \ldots, n\}$.

We have the following result:

Theorem. *In the chain ladder model of Schnaus, all chain ladder predictors are optimal under the sequential optimality criterion.*

Under the assumptions of the chain ladder model of Schnaus, the theorem provides a complete justification of the chain ladder method.

Sequential Conditional Linear Chain Ladder Model

The chain ladder model of Schnaus contains only assumptions on the conditional moments of first and second order and provides the basis for the formulation of the *sequential conditional linear chain ladder model:*

> **Sequential Conditional Linear Chain Ladder Model**: *There exist unknown parameters* $\varphi_1, \ldots, \varphi_n$ *and parameters* $\sigma_1^2, \ldots, \sigma_n^2$ *such that the identities*
>
> $$E(S_{i,k} \mid \mathcal{G}_{k-1}) = S_{i,k-1}\, \varphi_k$$
> $$\mathrm{cov}(S_{i,k}, S_{j,k} \mid \mathcal{G}_{k-1}) = S_{i,k-1}\, \sigma_k^2\, \delta_{i,j}$$
>
> *hold for every* $k \in \{1, \ldots, n\}$ *and for all* $i, j \in \{0, 1, \ldots, n-k+1\}$.

Thus, the sequential conditional linear chain ladder model contains, for every development year $k \in \{1, \ldots, n\}$, a conditional linear model and it can be shown that the chain ladder predictor $S_{n-k+1,k}^{\mathrm{CL}}$ is precisely the conditional Gauss–Markov predictor of $S_{n-k+1,k}$ in that model.

Comparison with Other Models

In the models considered so far, we have applied the statistical principle of minimizing the (conditional) expected squared prediction error.

Another statistical principle is the maximum likelihood principle, which also leads, for a wide class of *multinomial models*, to the chain ladder method. Therefore, it is interesting to compare these models:

> **Theorem.** *The special multinomial model with binomial, Poisson or negative binomial distributed ultimate losses is not compatible with the chain ladder model of Schnaus.*

As the *Poisson model* is just the multinomial model with Poisson distributed ultimate losses, the theorem implies that the Poisson model is not compatible with the chain ladder model of Schnaus. The Poisson model is a model with independent incremental losses and it can be shown in general that, with the exception of degenerate special cases, no model with independent incremental losses is compatible with the chain ladder model of Schnaus; see Hess and Schmidt (2002).

This example shows that different and even incompatible models combined with appropriate statistical principles of estimation or prediction may lead to the same method of loss reserving.

Remarks

In the original version of the chain ladder model of Schnaus it is assumed that the parameters are random variables; see Schmidt and Schnaus (1996). This assumption is insignificant as the parameters have to be estimated anyway, and it is also incompatible with the assumption of the existence of a development pattern; see Hess and Schmidt (2002).

The main difference between the chain ladder models of Mack and Schnaus consists in the substitution of the assumption of independent accident years by the assumption that the cumulative losses of the same development year are conditionally uncorrelated. This prepares the transition to the sequential conditional linear chain ladder model.

Notes

Keywords: Chain Ladder Method (Basics), Development Patterns (Basics), Development Patterns (Estimation), Linear Models (Basics), Marginal Sum Method, Multinomial Model, Multiplicative Models, Multivariate Methods, Poisson Model.

References: Hess & Schmidt [2002], Ludwig, Schmeißer & Thänert [2009], Mack [1993, 1994a, 1994b, 2002], Schmidt [1997, 1999a, 1999b, 2009, 2012], Schmidt & Schnaus [1996].

Chain Ladder Method (Prediction Error)

Klaus D. Schmidt

Consider the run-off square of cumulative losses:

Accident year	Development year						
	0	1	... k	... $n-i$	... $n-1$	n	
0	$S_{0,0}$	$S_{0,1}$	... $S_{0,k}$	... $S_{0,n-i}$	... $S_{0,n-1}$	$S_{0,n}$	
1	$S_{1,0}$	$S_{1,1}$	... $S_{1,k}$	... $S_{1,n-i}$	... $S_{1,n-1}$	$S_{1,n}$	
$\vdots$	$\vdots$	$\vdots$	$\vdots$	$\vdots$	$\vdots$	$\vdots$	
i	$S_{i,0}$	$S_{i,1}$	... $S_{i,k}$	... $S_{i,n-i}$	... $S_{i,n-1}$	$S_{i,n}$	
$\vdots$	$\vdots$	$\vdots$	$\vdots$	$\vdots$	$\vdots$	$\vdots$	
$n-k$	$S_{n-k,0}$	$S_{n-k,1}$	... $S_{n-k,k}$	... $S_{n-k,n-i}$	... $S_{n-k,n-1}$	$S_{n-k,n}$	
$\vdots$	$\vdots$	$\vdots$	$\vdots$	$\vdots$	$\vdots$	$\vdots$	
$n-1$	$S_{n-1,0}$	$S_{n-1,1}$	... $S_{n-1,k}$	... $S_{n-1,n-i}$	... $S_{n-1,n-1}$	$S_{n-1,n}$	
n	$S_{n,0}$	$S_{n,1}$	... $S_{n,k}$	... $S_{n,n-i}$	... $S_{n,n-1}$	$S_{n,n}$	

We assume that the cumulative losses $S_{i,k}$ are observable for $i + k \leq n$ and that they are non-observable for $i + k \geq n + 1$.

In this article we discuss the estimation of the conditional squared prediction errors of the chain ladder reserves under the assumptions of the chain ladder model of Mack.

K.D. Schmidt (✉)
Technische Universität Dresden, Dresden, Germany
e-mail: klaus.d.schmidt@tu-dresden.de

© Springer International Publishing Switzerland 2016
M. Radtke et al. (eds.), *Handbook on Loss Reserving*,
EAA Series, DOI 10.1007/978-3-319-30056-6_8

Chain Ladder Method

The *chain ladder method* predicts the future cumulative losses $S_{i,k}$ by the *chain ladder predictors*

$$S_{i,k}^{CL} := S_{i,n-i} \prod_{l=n-i+1}^{k} \varphi_l^{CL}$$

with the *chain ladder factors*

$$\varphi_k^{CL} := \frac{\sum_{j=0}^{n-k} S_{j,k}}{\sum_{j=0}^{n-k} S_{j,k-1}}$$

The chain ladder predictors of the cumulative losses yield the *chain ladder predictors*

$$Z_{i,k}^{CL} := \begin{cases} S_{i,n-i+1}^{CL} - S_{i,n-i} & \text{if } i + k = n + 1 \\ S_{i,k}^{CL} - S_{i,k-1}^{CL} & \text{else} \end{cases}$$

of the future incremental losses $Z_{i,k}$ with $i, k \in \{1, \ldots, n\}$ and $i + k \geq n + 1$ and then the *chain ladder predictors*

$$R_i^{CL} := \sum_{l=n-i+1}^{n} Z_{i,l}^{CL}$$

$$R_{(c)}^{CL} := \sum_{l=c-n}^{n} Z_{c-l,l}^{CL}$$

$$R^{CL} := \sum_{l=1}^{n} \sum_{j=n-l+1}^{n} Z_{j,l}^{CL}$$

of the accident year reserves R_i, with $i \in \{1, \ldots, n\}$, of the calendar year reserves $R_{(c)}$, with $c \in \{n+1, \ldots, 2n\}$, and of the aggregate loss reserve R. These reserve predictors are also called *chain ladder reserves*.

Chain Ladder Model of Mack

For $k \in \{0, 1, \ldots, n\}$ and $i \in \{0, 1, \ldots, n\}$, we denote by

$$\mathcal{F}_{i,k}$$

the σ-algebra generated by the family $\{S_{i,l}\}_{l\in\{0,1,\ldots,k\}}$. This σ-algebra contains the information from accident year i that is available at the end of development year k.

The chain ladder model of Mack consists of the following assumptions:

Chain Ladder Model of Mack:

(i) *The accident years are independent.*

(ii) *There exist unknown parameters $\varphi_1, \ldots, \varphi_n$ and parameters $\sigma_1^2, \ldots, \sigma_n^2$ such that the identities*

$$E(S_{i,k}\,|\,\mathcal{F}_{i,k-1}) = S_{i,k-1}\,\varphi_k$$
$$\mathrm{var}(S_{i,k}\,|\,\mathcal{F}_{i,k-1}) = S_{i,k-1}\,\sigma_k^2$$

hold for every $k \in \{1, \ldots, n\}$ and for all $i \in \{0, 1, \ldots, n\}$.

Prediction Errors and Estimators of the Prediction Errors

We denote by

$$\mathcal{D}$$

the σ-algebra generated by the family $\{S_{j,l}\}_{l\in\{0,1,\ldots,n\},\,j\in\{0,1,\ldots,n-l\}}$. This σ-algebra contains the information provided by the collection of all observable cumulative losses.

Under the assumptions of the chain ladder model of Mack, the conditional mean squared errors of prediction, given the σ-algebra $\mathcal{D}$, satisfy

$$
\begin{aligned}
&E\left(\left(R_i^{\mathrm{CL}} - R_i\right)^2 \,\middle|\, \mathcal{D}\right) \\
&= S_{i,n-i} \sum_{l=n-i+1}^{n} \left(\prod_{s=n-i+1}^{l-1} \varphi_s\right) \sigma_l^2 \left(\prod_{s=l+1}^{n} \varphi_s^2\right) + S_{i,n-i}^2 \left(\prod_{s=n-i+1}^{n} \varphi_s^{\mathrm{CL}} - \prod_{s=n-i+1}^{n} \varphi_s\right)^2
\end{aligned}
$$

for the chain ladder accident year reserves and

$$
\begin{aligned}
&E\left(\left(R^{\mathrm{CL}} - R\right)^2 \,\middle|\, \mathcal{D}\right) \\
&= \sum_{i=1}^{n} S_{i,n-i} \sum_{l=n-i+1}^{n} \left(\prod_{s=n-i+1}^{l-1} \varphi_s\right) \sigma_l^2 \left(\prod_{s=l+1}^{n} \varphi_s^2\right) \\
&\quad + \sum_{i=1}^{n}\sum_{j=1}^{n} S_{i,n-i} S_{j,n-j} \left(\prod_{s=n-i+1}^{n} \varphi_s^{\mathrm{CL}} - \prod_{s=n-i+1}^{n} \varphi_s\right)\left(\prod_{t=n-j+1}^{n} \varphi_t^{\mathrm{CL}} - \prod_{t=n-j+1}^{n} \varphi_t\right)
\end{aligned}
$$

for the chain ladder aggregate loss reserve.

In these identities, substitution of the variance parameters σ_k^2 by the estimators

$$\widehat{\sigma}_k^2 := \frac{1}{n-k} \sum_{j=0}^{n-k} S_{j,k-1} \left(\frac{S_{j,k}}{S_{j,k-1}} - \varphi_k^{\mathrm{CL}} \right)^2$$

for $k \in \{1, \ldots, n-1\}$ and by an estimator $\widehat{\sigma}_n^2$ obtained by extrapolation causes no problems. By contrast, substitution of the development factors φ_k by the chain ladder factors φ_k^{CL} would nullify the second summand.

To estimate the conditional mean squared errors of prediction, Mack (1993, 1994b, 2002) proposes the estimators

$$\widehat{E}\left(\left(R_i^{\mathrm{CL}} - R_i \right)^2 \,\Big|\, \mathcal{D} \right) := \left(S_{i,n}^{\mathrm{CL}} \right)^2 \sum_{l=n-i+1}^{n} \left(\frac{1}{\sum_{h=0}^{n-l} S_{h,l}} + \frac{1}{S_{i,l}^{\mathrm{CL}}} \right) \frac{\widehat{\sigma}_l^2}{\varphi_l^{\mathrm{CL}}}$$

and

$$\widehat{E}\left(\left(R^{\mathrm{CL}} - R \right)^2 \,\Big|\, \mathcal{D} \right)$$

$$:= \sum_{l=1}^{n} \sum_{i=n-l+1}^{n} \sum_{j=n-l+1}^{n} S_{i,n}^{\mathrm{CL}} S_{j,n}^{\mathrm{CL}} \left(\frac{1}{\sum_{h=0}^{n-l} S_{h,l}} + \frac{1}{S_{i,l}^{\mathrm{CL}}} \delta_{i,j} \right) \frac{\widehat{\sigma}_l^2}{\varphi_l^{\mathrm{CL}}}$$

The construction of these estimators is based on the assumptions of the chain ladder model of Mack and uses certain approximations.[1]

Remark

For a discussion of the prediction error of the chain ladder calendar year reserves we refer to Buchwalder et al. (2006).

Notes

Keywords: Chain Ladder Method (Basics), Chain Ladder Method (Models), Multivariate Methods.

References: Buchwalder, Bühlmann, Merz & Wüthrich [2006], Mack [1993, 1994b, 2002], Wüthrich & Merz [2008].

---

[1] These formulae are obtained if in the formulae of Mack the squared chain ladder factors are moved into the bracket and if it is taken into account that in the papers by Mack the indices of the parameters φ_k and σ_k^2 are shifted by 1 and the enumeration of accident and development years starts with 1 instead of 0.

Collective Model

Klaus D. Schmidt

The incremental or cumulative losses represented in a *run-off triangle* are either

- *numbers of claims* or
- *aggregate claim amounts*, which are defined as sums of *paid* or *incurred losses* of individual claims.

In the case of aggregate claim amounts,

- the number of claims over which summation extends is called the *claim number* and
- the paid or incurred loss attached to an individual claim is called the *claim size*.

There are at least two reasons which suggest representing an aggregate claim amount by means of the claim number and the claim sizes:

- The claim number is not affected by inflation.
- There is usually a substantial statistical basis for the claim sizes.

In stochastic modelling of an aggregate claim amount, the claim number and the claim sizes are assumed to be random variables.

The representation of an aggregate claim amount S by means of the claim number N and the claim sizes X_k is particularly convenient when the claim sizes are independent and identically distributed and also independent of the claim number. These are precisely the assumptions of the collective model of risk theory:

K.D. Schmidt (✉)
Technische Universität Dresden, Dresden, Germany
e-mail: klaus.d.schmidt@tu-dresden.de

© Springer International Publishing Switzerland 2016
M. Radtke et al. (eds.), *Handbook on Loss Reserving*,
EAA Series, DOI 10.1007/978-3-319-30056-6_9

Collective Model: *There exists a random variable N with values in $\mathbb{N}_0$ and an independent and identically distributed sequence $\{X_k\}_{k\in\mathbb{N}}$ of random variables which is independent of N such that the identity*

$$S = \sum_{k=1}^{N} X_k$$

holds.

In this article we assume that the assumptions of the collective model are fulfilled and we denote by X a random variable having the same distribution as every random variable of the sequence $\{X_k\}_{k\in\mathbb{N}}$.

Moments of the Aggregate Claim Amount

Because of Wald's identities the first and second moments of the aggregate claim amount are determined by the first and second moments of the claim number and the claim size:

Lemma (Wald's Identities). *The aggregate claim amount satisfies*

$$E[S] = E[N]\,E[X]$$

and

$$\mathrm{var}[S] = E[N]\,\mathrm{var}[X] + \mathrm{var}[N]\left(E[X]\right)^2$$

If the expected claim number $E[N]$ and the expected claim size $E[X]$ are replaced by estimators, then the first of Wald's identities suggests using the product of these estimators as an estimator of the expected aggregate claim amount $E[S]$. In the same way, an estimator of the variance of the aggregate claim amount can be obtained according to the second of Wald's identities.

Wald's identities allow a simple calculation of the expectation and the variance of the aggregate claim amount. These moments are of interest in Cantelli's inequality:

Lemma (Cantelli's Inequality). *The inequality*

$$P\big[\{S \geq E[S] + c\}\big] \leq \frac{\mathrm{var}[S]}{c^2 + \mathrm{var}[S]} < 1$$

holds for every $c \in (0, \infty)$.

For $\varepsilon \in (0, 1)$, Cantelli's inequality yields

$$P\left[\left\{S \geq E[S] + \sqrt{\frac{1-\varepsilon}{\varepsilon}\,\mathrm{var}[S]}\right\}\right] \leq \varepsilon$$

Therefore, Cantelli's inequality provides upper bounds for the *quantiles* of the distribution of the aggregate claim amount.

Example. For $\varepsilon := 0.10$, Cantelli's inequality yields

$$P\left[\left\{S \geq E[S] + 3\sqrt{\mathrm{var}[S]}\right\}\right] \leq 0.10$$

Thus the 90 % quantile of the distribution of the aggregate claim amount S is at most equal to $E[S] + 3\sqrt{\mathrm{var}[S]}$.

It should be noted that, depending on the distribution of the aggregate claim amount, the upper bound in Cantelli's inequality may be very rough.

Distribution of the Aggregate Claim Amount

Because of the identity

$$P[\{S \leq s\}] = \sum_{n=0}^{\infty} P[\{N = n\}] P\left[\left\{\sum_{k=1}^{n} X_k \leq s\right\}\right]$$

and the independence of the claim sizes, the distribution of the aggregate claim amount is determined by the distributions of the claim number and the claim size. However, the calculation of the distribution of the aggregate claim amount using the previous identity is in general quite time-consuming.

Under certain assumptions on the distributions of the claim number and the claim size, the distribution of the aggregate claim amount can be determined by recursion.

Consider first the claim number N and let

$$p_n := P[\{N = n\}]$$

The basis for Panjer's recursion is the following characterization of an important class of claim number distributions:

Theorem (Panjer Class). *The following properties are equivalent:*

(a) *There exist $a, b \in \mathbb{R}$ with $a + b > 0$ such that the identity*

$$p_n = \left(a + \frac{b}{n}\right) p_{n-1}$$

holds for all $n \in \mathbb{N}$.

(b) *The distribution of N is a binomial, Poisson or negative binomial distribution. In this case $a < 1$.*

The three claim number distributions that satisfy the recursion of the theorem are of different importance and differ in the relation between the variance and the expectation:

- The *binomial distribution* ($a < 0$) satisfies $\mathrm{var}[N] < E[N]$ and has minor importance as a claim number distribution.

- The *Poisson distribution* ($a = 0$) satisfies $\mathrm{var}[N] = E[N]$ and is particularly simple since it has only one parameter.
- The *negative binomial distribution* ($0 < a < 1$) satisfies $\mathrm{var}[N] > E[N]$ and can be represented as a mixed Poisson distribution with a mixing *Gamma distribution*. The negative binomial distribution often provides a better fit to the empirical claim number distribution than the Poisson distribution does.

These distributions are also of interest in the *Poisson model*, in the *multinomial model* and in the *credibility model of Witting*.

Consider now also the claim size X and the aggregate claim amount S and assume that $P[\{X \in \mathbb{N}_0\}] = 1$. Then we have $P[\{S \in \mathbb{N}_0\}] = 1$ as well. Let

$$f_n := P[\{X = n\}]$$
$$g_n := P[\{S = n\}]$$

The following theorem is the prototype of a series of results that lead from a recursion for the distribution of the claim number to a recursion for the distribution of the aggregate claim amount:

Theorem (Panjer's Recursion). *Assume that $f_0 = 0$. If there exist $a, b \in \mathbb{R}$ with $a + b > 0$ such that the identity*

$$p_n = \left(a + \frac{b}{n}\right) p_{n-1}$$

holds for all $n \in \mathbb{N}$, then $g_0 = p_0$ and the identity

$$g_n = \sum_{k=1}^{n} \left(a + k\,\frac{b}{n}\right) g_{n-k} f_k$$

holds for all $n \in \mathbb{N}$.

Under the same assumptions there is also a recursion for the moments of the aggregate claim amount, using moments of the claim number and the claim size.

Simulation

It depends on the distributions of the claim number and the claim size whether an explicit calculation of the distribution of the aggregate claim amount is possible or not. In the case where this is not possible, an approximation of the distribution of the aggregate claim amount can be obtained by *simulations* of the distributions of the claim number and the claim size.

Notes

Keywords: Credibility Models (Loss Reserving), Multinomial Model, Poisson Model, Run-Off Triangles, Simulation.

References: Hess [2009, 2012], Hess, Liewald & Schmidt [2002], Mack [2002], Schmidt [1996, 2009].

Controlling

Barbara Alfermann, Michael Radtke and Axel Reich

At least since the deregulation of the insurance markets and the globalization of the financial markets, target-oriented portfolio management has been a vital success factor for each insurance company. In a market environment made dynamic, the success of a company can only be ensured in the long term if the insurance companies move away from retrospective responses to an active value-oriented portfolio management. An anticipatory corporate policy in terms of a value-oriented management requires a permanent performance measurement for this purpose. The determination of real-time key performance indicators creates the necessary transparency in the insurance company and facilitates thus a risk- and return-oriented allocation of resources.

Actuarial controlling manages the core sector of the insurance business, the underwriting, target-oriented on the basis of ex-post data and ex-ante predictions. In this process the performance measurement ensures the permanent determination and analysis of key performance indicators in comparison with the original planning targets and other empirical values (*benchmarks*). However, the key performance indicators must not only be determined and analysed. After disclosing performance gaps, the primary objective is to develop corrective actions and to make precise agreements about these actions for an effective counteraction (*feedback*). If a course correction is not achievable, the planned targets have to be adjusted correspondingly (*feedforward*).

The entire controlling process can be represented as a classical control loop model with corresponding feedforward and feedback possibilities. Thus actuarial controlling is not the continuation of financial accounting by other means, but a forward-looking management approach. This approach is a starting point and return position for optimization impulses in the areas of portfolio management and premium policy as well as reserving, reinsurance and safety capital strategies and ties in directly

B. Alfermann · M. Radtke (✉) · A. Reich
Fachhochschule Dortmund, Dortmund, Germany
e-mail: michael.radtke@fh-dortmund.de

© Springer International Publishing Switzerland 2016
M. Radtke et al. (eds.), *Handbook on Loss Reserving*,
EAA Series, DOI 10.1007/978-3-319-30056-6_10

with the value generating business units in the insurance company. It ensures that operative actions lead to the achievement of the company's strategic objectives in terms of a sustainable accumulation of value. The accumulation of value of the insurance company is based on the free *cash flows* that are generable in the longer-term future, i.e. on the amount of achievable profit on a sustained basis. Its present value represents today's company value.

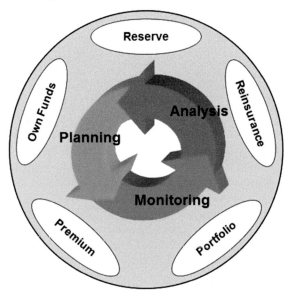

The Management Process as a Feedback Loop

The portfolio performance is the product of the materialising balance of risks between the transferred individual risks and hence, as the direct value driver, the real optimization variable of the insurance company. It has to be determined and evaluated with multi-criteria actuarial methods with regard to the profitability, the volatility and the growth. Not the individual risk, but the whole portfolio is controlled, because the portfolio performance results precisely from the interaction of numerous risks with the corresponding synergetic and compensatory effects.

Thereby homogeneity, balancing of risks and the size of the segment are definitely competing demands for the segmentation.

For example, the *homogeneity* as segmentation criterion is always a relative size: In the extreme case, the absolute homogeneous segment consists of the individual policyholder or the individual risk. Therewith the dilemma of any collective can be formulated as follows: The bigger the segment is, the more stable, but more heterogeneous it tends to be. A variety of proximity measures exists within the framework of multivariate methods for the measurement of similarities and distances between groups. With regard to the statistical significance of the loss experience, smaller segments are of course subject to stronger random influences because of their narrower basis for the balancing of risks and are accordingly more volatile (for example, due to the *large loss problem*).

According to this, the management approach for such a portfolio comprises essentially the loss costs and the portfolio.

The *loss costs*, representing the core element of the deliveries of an insurance company, are naturally of crucial importance and by far the largest cost block. However, compared to the operating or distribution costs they show peculiarities that require an actuarial treatment within the framework of a target-oriented controlling. Their stochastic character complicates not only the predictability, but also the performance measurement. The run-off problem represents another peculiarity of the loss costs, which results from a strong payment retardation, and also requires actuarial methods that ensure a reliable prediction of the ultimate losses within the framework of loss costs controlling.

Accident Year	0 > 1	1 > 2	2 > 3	
2005	1.360	1.049	1.024	
2006	1.372	1.048	1.028	
2007	1.351	1.047	1.022	
2008	1.361	1.049	1.030	
2009	1.330	1.048	1.028	**Calendar**
2010	1.321	1.042	1.023	**Year**
2011	1.340	1.046	1.028	
2012	1.313	1.044	1.026	
2013	1.310	1.042	1.028	
2014	1.309	1.042		
2015	1.297			
2016				
Mean	1.333	1.046	1.026	

Development Factors

A target-oriented loss costs controlling in the *middle-* and *long-tail* business has to give particular attention to the problem of late reported losses. The number of late reported losses as well as the amount of these future contractual obligations can only ultimately be identified with a huge time delay after their occurrence. Therefore the question of which run-off potential emanates from a portfolio has to be answered by a reliable prediction of the ultimate losses and by the resulting determination of risk-adequate reserves within the framework of an analysis on accident respectively underwriting year view. The accident year view overcomes the inertia of the calendar year view, which mixes all run-off effects of previous accident years and makes it difficult to recognize trends and structural breaks. After the estimation of the complete run-off structure, the accident year results, which are decision-relevant from an underwriting perspective, can be easily converted into the corresponding calendar year view, which is reflected in the balance sheet.

Naturally in the middle- and long-tail business, the *risk of change* plays a substantial role besides the risk of random fluctuation. Therefore inflation-related changes as well as changing demand behaviour of policyholders or changes in claims handling have to be taken into account within the framework of late reported loss predictions.

The question of changing *development speeds* has to be separated clearly from trends in *loss ratios* and *premium rates*. Thereby not only the numerical sensitivity of the original data has to be considered appropriately. *Large loss smoothing*, trend analyses on gliding time horizons and the comparison with *development patterns* from external market statistics indicate the spectrum of the used toolkit and the variety of techniques.

Separate analyses for the number of claims and for the claim size improve the quality of results by isolating the IBNER effect (*incurred but not enough reserved*) from the IBNR effect (*incurred but not reported*).

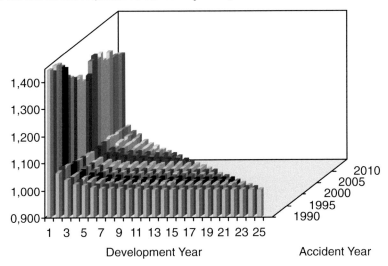

Faster Run-Off: Decreasing Development Factors

The early identification of development trends and a permanent adaption of planning targets to the recent developments are important for the responsiveness in the current market environment.

Within the framework of a sensitive and fast responding portfolio management, the differentiated analysis of the *run-off triangles* is decisive for the detection of *early warning indicators* and is therewith a precondition for an effective loss costs controlling. Subsequently the quality of the analyses and predictions can be reconciled over time for each calendar year by means of comparisons of *actual versus expected* losses.

The strong payment retardation in middle- and long-tail business generates a *cash flow* from the premium paid in advance, on which *interests* are gained. Therefore, following the determination of the nominal ultimate loss ratios, the integration of these cash flow effects is necessary for the performance measurement: The present values of the premiums and particularly the present values of the losses have to be calculated. Hence the amounts that would be payable, if the future payments were due right at the beginning of the accident year, are determined. This translation of the ultimate losses to an earlier date by means of discounting enables the comparison of discounted *combined ratios* or of discounted business results with the target and minimum returns of the insurance company. Obviously, the expected returns set for single business segments should reflect the fluctuation potential of these segments.

The risk- and return-oriented comparison of the target objectives with the realized performance reveals, whether and to what extent the portfolio is currently above the *break-even* point, how stable its profit situation and how target-oriented its development is and to what extend the *return on equity* targets set under *shareholder value* aspects can be realized. The corresponding break-even and target loss ratios can be determined for the underwriting. Within the framework of a company-wide management approach, a *scoring* of all business segments can then be performed on this basis and a corresponding catalogue of measures can be derived.

Last but not least, the reserve quality and hence the reserving policy in the past are future competition parameters. Since the effects of very different accident years are realized in a calendar year (the current calendar year corresponds to the main diagonal of the run-off triangle), the present reserve quality also affects future calendar years.

Beyond the quantification of the extent of over- or under-reserving at a specific point in time, of course a multi-year comparison is decisive for the interpretation of results: Thus the changes compared to the required reserve analysis from the previous year can be separated for each segment into effects from changes of the model parameters, the influence of the main diagonal and currency effects, for example.

In an increasingly competitive environment the financing of the calendar year loss ratios by the run-off results of the past is more and more relevant. Only prediction of the ultimate losses can determine to what extent the previous reserving policy takes into account such strategic options or whether a narrowed room for manoeuvre arises compared to other competitors. The cushions and inherited obligations of a portfolio are a decisive influential factor considering the question: *How well prepared is the insurance company for the price competition in the market?*

Notes

Keywords: Development Patterns (Basics), Run-Off Triangles.

Credibility Models (Basics)

Klaus D. Schmidt

Credibility methods predict a non-observable random variable or a non-observable random vector by an affine-linear transformation of observable random variables which satisfies a certain optimality criterion.

In this article we confine ourselves to the simplest and also most familiar optimality criterion, which is also used for prediction in a *linear model*. This optimality criterion requires minimization of the expected squared prediction error and has the advantage that it is based only on the knowledge of the first and second order moments of the random variables.

Most models in which a credibility method is applied also involve a random risk parameter, which in many cases has a very natural interpretation. These models typically include assumptions on the distribution of the risk parameter as well as assumptions on the conditional joint distribution of all observable and non-observable random variables, given the risk parameter, such that the required first and second order moments can be determined from these assumptions.

Credibility models with a risk parameter have their origin in pricing and are also used in loss reserving.

Elementary Credibility Model

In this section we consider random variables $X_1, \ldots, X_n$ and X_0, and we assume that $X_1, \ldots, X_n$ are observable and that X_0 is non-observable. We study the problem of optimal prediction of X_0 by an affine-linear function of $X_1, \ldots, X_n$.

K.D. Schmidt (✉)
Technische Universität Dresden, Dresden, Germany
e-mail: klaus.d.schmidt@tu-dresden.de

© Springer International Publishing Switzerland 2016
M. Radtke et al. (eds.), *Handbook on Loss Reserving*,
EAA Series, DOI 10.1007/978-3-319-30056-6_11

In principle, every random variable $\widehat{Y}$ which depends on hazard only via the observable random variables $X_1, \ldots, X_n$ is a possible *predictor* of X_0. For a predictor $\widehat{Y}$ of X_0, the difference

$$\widehat{Y} - X_0$$

is called the *prediction error* of $\widehat{Y}$ and the expectation

$$E\left[\left(\widehat{Y} - X_0\right)^2\right]$$

is called the *expected squared prediction error* of $\widehat{Y}$. The expected squared prediction error provides the basis for the comparison of different predictors of X_0. Because of the identity

$$E\left[\left(\widehat{Y} - X_0\right)^2\right] = \mathrm{var}\left[\widehat{Y} - X_0\right] + \left(E\left[\widehat{Y} - X_0\right]\right)^2$$

the expected squared prediction error of a predictor of X_0 is determined by the expectation and the variance of the prediction error.

We consider now some possible properties of a predictor of X_0: A predictor $\widehat{Y}$ is called

- an *affine-linear predictor* of X_0 if there exist real numbers $a, a_1, \ldots, a_n$ such that

$$\widehat{Y} = a + \sum_{k=1}^{n} a_k\, X_k$$

- a *credibility predictor* of X_0 if it is an affine-linear predictor of X_0 which minimizes the expected squared prediction error over all affine-linear predictors of X_0;
- an *unbiased predictor* of X_0 if

$$E\left[\widehat{Y}\right] = E[X_0]$$

For every unbiased predictor $\widehat{Y}$ of X_0 the expected squared prediction error satisfies

$$E\left[\left(\widehat{Y} - X_0\right)^2\right] = \mathrm{var}\left[\widehat{Y} - X_0\right]$$

Therefore, the expected squared prediction error of an unbiased predictor of X_0 is determined by the variance of the prediction error.

It follows from the Hilbert space projection theorem that there exists a unique credibility predictor

$$X_0^{\mathrm{CR}}$$

of X_0 and that the credibility predictor of X_0 is an unbiased predictor of X_0. Unbiasedness of the credibility predictor of X_0 is remarkable, since it is not required in the definition of a credibility predictor of X_0. Under an additional assumption on the joint distribution of the observable random variables, the existence and uniqueness of the credibility predictor can also be proven without using of the projection theorem.

Define now

$$\mathbf{X}_1 := \begin{pmatrix} X_1 \\ \vdots \\ X_n \end{pmatrix}$$

and

$$\begin{aligned} \boldsymbol{\mu}_1 &:= E[\mathbf{X}_1] \\ \Sigma_{11} &:= \mathrm{var}[\mathbf{X}_1] \\ \boldsymbol{\varrho}_{10} &:= \mathrm{cov}[\mathbf{X}_1, X_0] \\ \mu_0 &:= E[X_0] \\ \sigma_0^2 &:= \mathrm{var}[X_0] \end{aligned}$$

Then we have

$$E\left[\begin{pmatrix} \mathbf{X}_1 \\ X_0 \end{pmatrix}\right] = \left[\begin{pmatrix} \boldsymbol{\mu}_1 \\ \mu_0 \end{pmatrix}\right]$$

and

$$\mathrm{var}\left[\begin{pmatrix} \mathbf{X}_1 \\ X_0 \end{pmatrix}\right] = \left[\begin{pmatrix} \Sigma_{11} & \boldsymbol{\varrho}_{10} \\ \boldsymbol{\varrho}'_{10} & \sigma_0^2 \end{pmatrix}\right]$$

and every affine-linear predictor of X_0 has the form

$$\widehat{Y} = a + \mathbf{a}'\mathbf{X}_1$$

with a real number a and a vector $\mathbf{a}$. The following result can be proven by means of differential calculus, for example:

Theorem. *Assume that Σ_{11} is invertible. Then there exists a unique credibility predictor X_0^{CR} of X_0 and it satisfies*

$$X_0^{\mathrm{CR}} = \mu_0 + \boldsymbol{\varrho}'_{10}\,\Sigma_{11}^{-1}(\mathbf{X}_1 - \boldsymbol{\mu}_1)$$

and

$$E\left[(X_0^{\mathrm{CR}} - X_0)^2\right] = \sigma_0^2 - \boldsymbol{\varrho}'_{10}\,\Sigma_{11}^{-1}\,\boldsymbol{\varrho}_{10}$$

In particular, the credibility predictor of X_0 is an unbiased predictor of X_0.

The theorem yields an interesting corollary:

Corollary. *Assume that* Σ_{11} *is invertible and that* $\varrho_{10} = \mathbf{0}$. *Then*

$$X_0^{\text{CR}} = \mu_0$$

Thus, under the assumptions of the corollary, the credibility predictor of X_0 does not depend on any of the observable random variables.

Standard Credibility Model

The formula for the credibility predictor involves the inverse of the variance of the vector of observable random variables. We consider now a family of credibility models in which the variance of the vector of observable random variables is invertible and its inverse can be given in closed form.

Standard Credibility Model: *There exist parameters* μ *and* λ *as well as* $\varphi_0, \varphi_1, \ldots,$ $\varphi_n \in (0, \infty)$ *with*

$$1 + \lambda \sum_{k=1}^{n} \varphi_k^{-1} > 0$$

such that the identities

$$E[X_k] = \mu$$
$$\text{cov}[X_k, X_l] = \varphi_k\, \delta_{k,l} + \lambda$$

hold for all $k, l \in \{0, 1, \ldots, n\}$.

Define

$$\mathbf{1} := \begin{pmatrix} 1 \\ \vdots \\ 1 \end{pmatrix}$$

and

$$\mathbf{E} := \mathbf{1}\mathbf{1}'$$

Then the standard credibility model can be formulated as follows:

Standard Credibility Model: *There exist parameters* μ *and* λ *as well as* $\varphi_0 \in (0, \infty)$ *and a positive definite diagonal matrix* Φ *with*

$$1 + \lambda\, \mathbf{1}'\Phi^{-1}\mathbf{1} > 0$$

such that

$$\boldsymbol{\mu}_1 = \mu\,\mathbf{1}$$
$$\Sigma_{11} = \Phi + \lambda\,\mathbf{E}$$
$$\varrho_{10} = \lambda\,\mathbf{1}$$
$$\mu_0 = \mu$$
$$\sigma_0^2 = \varphi_0 + \lambda$$

We obtain the following result:

Lemma. *Under the assumptions of the standard credibility model, the matrix Σ_{11} is invertible and its inverse satisfies*

$$\Sigma_{11}^{-1} = \Phi^{-1} - \frac{\lambda}{1 + \lambda\,\mathbf{1}'\Phi^{-1}\mathbf{1}}\,\Phi^{-1}\mathbf{E}\Phi^{-1}$$

Under the assumptions of the standard credibility model we can determine the credibility predictor of X_0 and its expected squared prediction error:

Theorem. *Under the assumptions of the standard credibility model, the credibility predictor X_0^{CR} of X_0 satisfies*

$$X_0^{\mathrm{CR}} = \frac{1}{1 + \lambda\sum_{l=1}^{n}\varphi_l^{-1}}\,\mu + \sum_{k=1}^{n}\frac{\lambda\,\varphi_k^{-1}}{1 + \lambda\sum_{l=1}^{n}\varphi_l^{-1}}\,X_k$$

and

$$E\big[(X_0^{\mathrm{CR}} - X_0)^2\big] = \varphi_0 + \frac{1}{1 + \lambda\sum_{l=1}^{n}\varphi_l^{-1}}\,\lambda$$

Thus, under the assumptions of the standard credibility model, the credibility predictor of X_0 is a weighted mean of the expectation μ and the observable random variables $X_1, \ldots, X_n$.

Extended Credibility Model

Consider now a random vector $\mathbf{X}$ with values in $\mathbb{R}^m$ of which the first m_1 coordinates are observable and the last $m_2 = m - m_1$ coordinates are non-observable. We write $\mathbf{X}$ as a block vector

$$\mathbf{X} = \begin{pmatrix} \mathbf{X}_1 \\ \mathbf{X}_2 \end{pmatrix}$$

and we assume that $\mathbf{X}_1$ is observable and that $\mathbf{X}_2$ is non-observable. Letting

$$\boldsymbol{\mu}_1 := E[\mathbf{X}_1]$$
$$\boldsymbol{\mu}_2 := E[\mathbf{X}_2]$$

and

$$\Sigma_{11} := \text{var}[\mathbf{X}_1]$$
$$\Sigma_{12} := \text{cov}[\mathbf{X}_1, \mathbf{X}_2]$$
$$\Sigma_{21} := \text{cov}[\mathbf{X}_2, \mathbf{X}_1]$$
$$\Sigma_{22} := \text{var}[\mathbf{X}_2]$$

we obtain

$$E[\mathbf{X}] = \begin{pmatrix} \mu_1 \\ \mu_2 \end{pmatrix}$$

and

$$\text{var}[\mathbf{X}] = \begin{pmatrix} \Sigma_{11} & \Sigma_{12} \\ \Sigma_{21} & \Sigma_{22} \end{pmatrix}$$

We study the general problem of optimal prediction of $\mathbf{DX}_2$ with a matrix $\mathbf{D} \in \mathbb{R}^{r \times m_2}$.

The problem of optimal prediction of $\mathbf{DX}_2$ contains the following special cases: Optimal prediction of $\mathbf{X}_2$, optimal prediction of a linear combination of the coordinates of $\mathbf{X}_2$ and optimal prediction of a single coordinate of $\mathbf{X}_2$.

In principle, every random vector $\widehat{\mathbf{Y}}$ with values in $\mathbb{R}^r$ which depends on hazard only via the random vector $\mathbf{X}_1$ is a possible *predictor* of $\mathbf{DX}_2$. For a predictor $\widehat{\mathbf{Y}}$ of $\mathbf{DX}_2$, the difference

$$\widehat{\mathbf{Y}} - \mathbf{DX}_2$$

is called the *prediction error* of $\widehat{\mathbf{Y}}$ and the expectation

$$E\left[(\widehat{\mathbf{Y}} - \mathbf{DX}_2)'(\widehat{\mathbf{Y}} - \mathbf{DX}_2)\right]$$

is called the *expected squared prediction error* of $\widehat{\mathbf{Y}}$. The expected squared prediction error provides the basis for the comparison of different predictors of $\mathbf{DX}_2$. Because of the identity

$$E\left[(\widehat{\mathbf{Y}} - \mathbf{DX}_2)'(\widehat{\mathbf{Y}} - \mathbf{DX}_2)\right]$$
$$= \text{trace}\left(\text{var}[\widehat{\mathbf{Y}} - \mathbf{DX}_2]\right) + E[\widehat{\mathbf{Y}} - \mathbf{DX}_2]' E[\widehat{\mathbf{Y}} - \mathbf{DX}_2]$$

the expected squared prediction error of a predictor $\widehat{\mathbf{Y}}$ of $\mathbf{DX}_2$ is determined by the expectation and the variance of the prediction error.

We consider now some possible properties of a predictor of $\mathbf{DX}_2$: A predictor $\widehat{\mathbf{Y}}$ with values in $\mathbb{R}^r$ is called

- an *affine-linear predictor* of $\mathbf{DX}_2$ if there exist a vector $\mathbf{q}$ and a matrix $\mathbf{Q}$ such that

$$\widehat{\mathbf{Y}} = \mathbf{q} + \mathbf{Q}\mathbf{X}_1$$

- a *credibility predictor* of $\mathbf{DX}_2$ if it is an affine-linear predictor of $\mathbf{DX}_2$ which minimizes the expected squared prediction error over all affine-linear predictors of $\mathbf{DX}_2$;
- an *unbiased predictor* of $\mathbf{DX}_2$ if

$$E\left[\widehat{\mathbf{Y}}\right] = E[\mathbf{DX}_2]$$

Every unbiased predictor $\widehat{\mathbf{Y}}$ of $\mathbf{DX}_2$ satisfies

$$E\left[\left(\widehat{\mathbf{Y}} - \mathbf{DX}_2\right)'\left(\widehat{\mathbf{Y}} - \mathbf{DX}_2\right)\right] = \mathrm{trace}\left(\mathrm{var}\left[\widehat{\mathbf{Y}} - \mathbf{DX}_2\right]\right)$$

Therefore, the expected squared prediction error of an unbiased predictor $\widehat{\mathbf{Y}}$ of $\mathbf{DX}_2$ is determined by the variance of the prediction error.

Theorem. *Assume that Σ_{11} is invertible. Then there exists a unique credibility predictor $(\mathbf{DX}_2)^{\mathrm{CR}}$ of $\mathbf{DX}_2$ and it satisfies*

$$(\mathbf{DX}_2)^{\mathrm{CR}} = \mathbf{D}\left(\mu_2 + \Sigma_{21}\Sigma_{11}^{-1}(\mathbf{X}_1 - \mu_1)\right)$$

and

$$\mathrm{var}\left[(\mathbf{DX}_2)^{\mathrm{CR}} - \mathbf{DX}_2\right] = \mathbf{D}\left(\Sigma_{22} - \Sigma_{21}\Sigma_{11}^{-1}\Sigma_{12}\right)\mathbf{D}'$$

In particular, the credibility predictor of $\mathbf{DX}_2$ is an unbiased predictor of $\mathbf{DX}_2$.

The theorem yields an interesting corollary:

Corollary. *Assume that Σ_{11} is invertible. Then*

$$\mathbf{X}_2^{\mathrm{CR}} = \mu_2 + \Sigma_{21}\Sigma_{11}^{-1}(\mathbf{X}_1 - \mu_1)$$

and

$$(\mathbf{DX}_2)^{\mathrm{CR}} = \mathbf{DX}_2^{\mathrm{CR}}$$

The last identity of the corollary is called *linearity* of credibility prediction. In particular, the corollary shows that the credibility predictor of

$$\mathbf{X}_2 = \begin{pmatrix} X_{m_1+1} \\ \vdots \\ X_m \end{pmatrix}$$

satisfies

$$\mathbf{X}_2^{\mathrm{CR}} = \begin{pmatrix} X_{m_1+1}^{\mathrm{CR}} \\ \vdots \\ X_m^{\mathrm{CR}} \end{pmatrix}$$

Therefore, the coordinates of the credibility predictor of $\mathbf{X}_2$ are precisely the credibility predictors of the coordinates of $\mathbf{X}_2$.

We conclude this section with a special case:

Corollary. *Assume that Σ_{11} is invertible and that $\Sigma_{21} = \mathbf{O}$. Then*

$$\mathbf{X}_2^{CR} = \mu_2$$

and

$$\mathrm{var}\big[\mathbf{X}_2^{CR} - \mathbf{X}_2\big] = \Sigma_{22}$$

Thus, under the assumptions of the corollary, the credibility predictor of $\mathbf{X}_2$ does not depend on any of the observable random variables.

Reduction of the Observable Random Variables

In this section we consider a random vector

$$\mathbf{X} = \begin{pmatrix} \mathbf{X}_1 \\ \mathbf{X}_2 \\ \mathbf{X}_3 \end{pmatrix}$$

and we assume that $\mathbf{X}_1$ and $\mathbf{X}_3$ are observable and that $\mathbf{X}_2$ is non-observable. Letting

$$\mu_i := E[\mathbf{X}_i]$$

and

$$\Sigma_{ik} := \mathrm{cov}[\mathbf{X}_i, \mathbf{X}_k]$$

for $i, k \in \{1, 2, 3\}$ we obtain

$$E[\mathbf{X}] = \begin{pmatrix} \mu_1 \\ \mu_2 \\ \mu_3 \end{pmatrix}$$

and

$$\mathrm{var}[\mathbf{X}] = \begin{pmatrix} \Sigma_{11} & \Sigma_{12} & \Sigma_{13} \\ \Sigma_{21} & \Sigma_{22} & \Sigma_{23} \\ \Sigma_{31} & \Sigma_{32} & \Sigma_{33} \end{pmatrix}$$

If the matrix

$$\begin{pmatrix} \Sigma_{11} & \Sigma_{13} \\ \Sigma_{31} & \Sigma_{33} \end{pmatrix}$$

is invertible, then the credibility predictor of $\mathbf{X}_2$ on the basis of $\mathbf{X}_1$ and $\mathbf{X}_3$ satisfies

$$\mathbf{X}_2^{CR} = \mu_2 + \left(\Sigma_{21}\ \Sigma_{23}\right) \begin{pmatrix} \Sigma_{11} & \Sigma_{13} \\ \Sigma_{31} & \Sigma_{33} \end{pmatrix}^{-1} \begin{pmatrix} \mathbf{X}_1 - \mu_1 \\ \mathbf{X}_3 - \mu_3 \end{pmatrix}$$

The following theorem shows that under certain conditions this formula can be simplified:

Theorem. *Assume that Σ_{11} is invertible and that $\Sigma_{13} = \mathbf{O}$ and $\Sigma_{23} = \mathbf{O}$. Then*

$$\mathbf{X}_2^{CR} = \mu_2 + \Sigma_{21}\Sigma_{11}^{-1}(\mathbf{X}_1 - \mu_1)$$

The theorem asserts that a group of observable random variables which are uncorrelated with all other (observable or non-observable) random variables of the model (and form the random vector $\mathbf{X}_3$ used in the theorem) can be disregarded in the determination of the credibility predictor of $\mathbf{X}_2$ and therefore can be eliminated from the model from the very beginning.

Credibility Models with a Risk Parameter

In credibility models the assumptions on the distribution of the random vector $\mathbf{X}$ of all observable and non-observable random variables are usually made implicitly by formulating assumptions on the distribution of a non-observable *risk parameter* Λ as well as assumptions on the conditional distribution of $\mathbf{X}$ given Λ. The risk parameter Λ can be a random variable or a random vector.

In credibility models in which the expected squared prediction error has to be minimized, only the first and second order moments of $\mathbf{X}$ are needed. Because of the identities

$$E[\mathbf{X}] = E[E(\mathbf{X}|\Lambda)]$$
$$\text{var}[\mathbf{X}] = E[\text{var}(\mathbf{X}|\Lambda)] + \text{var}[E(\mathbf{X}|\Lambda)]$$

it is sufficient to formulate

- assumptions on the conditional moments $E(\mathbf{X}|\Lambda)$ and $\text{var}(\mathbf{X}|\Lambda)$ of $\mathbf{X}$ given Λ and
- assumptions on the unconditional moments $E[E(\mathbf{X}|\Lambda)]$, $\text{var}[E(\mathbf{X}|\Lambda)]$ and $E[\text{var}(\mathbf{X}|\Lambda)]$.

Since $E(\mathbf{X}|\Lambda)$ and $\text{var}(\mathbf{X}|\Lambda)$ are vector-valued respectively matrix-valued transformations of Λ, the second group of assumptions actually concerns the distribution of Λ.

Other Loss Functions

Besides the expected squared prediction error

$$E\big[(\widehat{Y} - X_0)^2\big]$$

considered here, occasionally other real-valued functions of the prediction error $\widehat{Y} - X_0$ are also used as an optimality criterion.

An example is the *expected weighted squared prediction error*

$$E\big[(\widehat{Y} - X_0)^2\, W\big]$$

with a strictly positive random variable W. With an appropriate choice of W and under certain assumptions on the joint distribution of W and $X_0, X_1, \ldots, X_n$, it is possible to ensure that the expectation of the credibility predictor of X_0 determined under the expected weighted squared prediction error is at least as large as that of X_0 and hence contains a *safety margin*. Details can be found in Schmidt (1992) and in Schmidt and Timpel (1995).

Notes

Keywords: Credibility Models (Loss Reserving), Linear Models (Basics).

References: Hess & Schmidt [2001], Schmidt [1992, 2012], Schmidt & Timpel [1995].

Credibility Models (Loss Reserving)

Klaus Th. Hess and Klaus D. Schmidt

Credibility models and credibility predictors are also useful in loss reserving. Some particularities arise from the structure of the run-off square and from the task to determine credibility predictors of different reserves. We explain these particularities and then discuss three credibility models for loss reserving.

Consider the run-off square of incremental losses:

Accident year	Development year					
	0	1	... k	... $n-i$	... $n-1$	n
0	$Z_{0,0}$	$Z_{0,1}$	... $Z_{0,k}$	... $Z_{0,n-i}$	... $Z_{0,n-1}$	$Z_{0,n}$
1	$Z_{1,0}$	$Z_{1,1}$	... $Z_{1,k}$	... $Z_{1,n-i}$	... $Z_{1,n-1}$	$Z_{1,n}$
$\vdots$	$\vdots$	$\vdots$	$\vdots$	$\vdots$	$\vdots$	$\vdots$
i	$Z_{i,0}$	$Z_{i,1}$	... $Z_{i,k}$	... $Z_{i,n-i}$	... $Z_{i,n-1}$	$Z_{i,n}$
$\vdots$	$\vdots$	$\vdots$	$\vdots$	$\vdots$	$\vdots$	$\vdots$
$n-k$	$Z_{n-k,0}$	$Z_{n-k,1}$	... $Z_{n-k,k}$	... $Z_{n-k,n-i}$	... $Z_{n-k,n-1}$	$Z_{n-k,n}$
$\vdots$	$\vdots$	$\vdots$	$\vdots$	$\vdots$	$\vdots$	$\vdots$
$n-1$	$Z_{n-1,0}$	$Z_{n-1,1}$	... $Z_{n-1,k}$	... $Z_{n-1,n-i}$	... $Z_{n-1,n-1}$	$Z_{n-1,n}$
n	$Z_{n,0}$	$Z_{n,1}$	... $Z_{n,k}$	... $Z_{n,n-i}$	... $Z_{n,n-1}$	$Z_{n,n}$

We assume that the incremental losses $Z_{i,k}$ are observable for $i + k \leq n$ and that they are non-observable for $i + k \geq n + 1$. For $i, k \in \{0, 1, \ldots, n\}$ we denote by

K.Th. Hess
Universität Rostock, Rostock, Germany
e-mail: klaus-thomas.hess@uni-rostock.de

K.D. Schmidt (✉)
Technische Universität Dresden, Dresden, Germany
e-mail: klaus.d.schmidt@tu-dresden.de

© Springer International Publishing Switzerland 2016
M. Radtke et al. (eds.), *Handbook on Loss Reserving*,
EAA Series, DOI 10.1007/978-3-319-30056-6_12

$$S_{i,k} := \sum_{l=0}^{k} Z_{i,l}$$

the cumulative loss from accident year i in development year k. We represent the incremental losses by a random vector

$$\mathbf{Z} = \begin{pmatrix} \mathbf{Z}_1 \\ \mathbf{Z}_2 \end{pmatrix}$$

such that $\mathbf{Z}_1$ consists of the $(n+1)(n+2)/2$ observable incremental losses and $\mathbf{Z}_2$ consists of the $n(n+1)/2$ non-observable incremental losses.

We are interested in

- the credibility predictors

$$\left(\sum_{l=n-i+1}^{n} Z_{i,l} \right)^{\mathrm{CR}}$$

of the accident year reserves $\sum_{l=n-i+1}^{n} Z_{i,l}$ with $i \in \{1, \dots, n\}$,
- the credibility predictors

$$\left(\sum_{l=c-n}^{n} Z_{c-l,l} \right)^{\mathrm{CR}}$$

of the calendar year reserves $\sum_{l=c-n}^{n} Z_{c-l,l}$ with $c \in \{n+1, \dots, 2n\}$, and
- the credibility predictor

$$\left(\sum_{l=1}^{n} \sum_{j=n-l+1}^{n} Z_{j,l} \right)^{\mathrm{CR}}$$

of the aggregate loss reserve $\sum_{l=1}^{n} \sum_{j=n-l+1}^{n} Z_{j,l}$.

These credibility predictors are also called *credibility reserves*.

In all three cases, one has to determine the credibility predictor of a random variable of the form $\mathbf{d}'\mathbf{Z}_2$ with a vector $\mathbf{d}$ of appropriate dimension, and it follows from the general results of credibility theory that the credibility predictor of $\mathbf{d}'\mathbf{Z}_2$ satisfies

$$(\mathbf{d}'\mathbf{Z}_2)^{\mathrm{CR}} = \mathbf{d}'\mathbf{Z}_2^{\mathrm{CR}}$$

Since the coordinates of the credibility predictor of $\mathbf{Z}_2$ are equivalent to the credibility predictors of the coordinates of $\mathbf{Z}_2$, we obtain

$$\left(\sum_{l=n-i+1}^{n} Z_{i,l} \right)^{\mathrm{CR}} = \sum_{l=n-i+1}^{n} Z_{i,l}^{\mathrm{CR}}$$

$$\left(\sum_{l=c-n}^{n} Z_{c-l,l} \right)^{\text{CR}} = \sum_{l=c-n}^{n} Z_{c-l,l}^{\text{CR}}$$

$$\left(\sum_{l=1}^{n} \sum_{j=n-l+1}^{n} Z_{j,l} \right)^{\text{CR}} = \sum_{l=1}^{n} \sum_{j=n-l+1}^{n} Z_{j,l}^{\text{CR}}$$

Therefore, the credibility reserves are precisely the sums of the credibility predictors of the corresponding incremental losses.

To predict a single non-observable incremental loss $Z_{i,s}$ it is sufficient to consider the model

$$\mathbf{Z}_{i,s} := \begin{pmatrix} \mathbf{Z}_1 \\ Z_{i,s} \end{pmatrix}$$

with $s \in \{n-i+1, \dots, n\}$.

The three credibility models considered below share two properties, which lead to a reduction of the prediction problem:

- In each of these credibility models, it is assumed that the incremental losses from different accident years are uncorrelated. This implies that the credibility predictor of a non-observable incremental loss $Z_{i,s}$ depends only on the observable incremental losses from accident year i such that it is sufficient to consider the model

$$\mathbf{Z}_{i,s}^{(i)} := \begin{pmatrix} \mathbf{Z}_1^{(i)} \\ Z_{i,s} \end{pmatrix}$$

where $\mathbf{Z}_1^{(i)}$ is the random vector consisting of the observable incremental losses $Z_{i,0}, Z_{i,1}, \dots, Z_{i,n-i}$ from accident year i. For this model we use the simplified notation

$$\{Z_{i,0}, Z_{i,1}, \dots, Z_{i,n-i}, Z_{i,s}\}$$

- In each of these credibility models, it turns out that the model

$$\{Z_{i,0}, Z_{i,1}, \dots, Z_{i,n-i}, Z_{i,s}\}$$

can be transformed into a model

$$\{X_{i,0}, X_{i,1}, \dots, X_{i,n-i}, X_{i,s}\}$$

by the transition to appropriately scaled incremental losses

$$X_{i,k} := Z_{i,k}/c_{i,k}$$

The latter model fulfills the assumptions of the *standard credibility model*, which in turn yields a general formula for the credibility predictor of $X_{i,s}$. Because of the identity

$$Z_{i,s}^{\mathrm{CR}} = (c_{i,s} X_{i,s})^{\mathrm{CR}} = c_{i,s} X_{i,s}^{\mathrm{CR}}$$

we then obtain a formula for the credibility predictor of $Z_{i,s}$.

Thus, the identification of the credibility reserves in the credibility models for loss reserving considered here uses several general properties of credibility predictors.

Credibility Model of Mack

The credibility model of Mack consists of the following assumptions:

Credibility Model of Mack:

(i) *The incremental losses from different accident years are uncorrelated.*

(ii) *There exist parameters $\zeta_0, \zeta_1, \ldots, \zeta_n$ and random variables $\Lambda_0, \Lambda_1, \ldots, \Lambda_n$ as well as functions $m_0, m_1, \ldots, m_n$ and $v_{0,0}, v_{0,1}, \ldots, v_{n,n}$ such that the identities*

$$E(Z_{i,k}|\Lambda_i) = m_i(\Lambda_i)\,\zeta_k$$
$$\mathrm{cov}(Z_{i,k}, Z_{i,l}|\Lambda_i) = v_{i,k}(\Lambda_i)\,\delta_{k,l}$$

hold for all $i, k, l \in \{0, 1, \ldots, n\}$.

The credibility model of Mack generalizes the *credibility model of De Vylder*.

We assume in this section that the assumptions of the credibility model of Mack are fulfilled.

Define

$$\alpha_i := E[m_i(\Lambda_i)] \sum_{l=0}^{n} \zeta_l$$

and

$$\vartheta_k := \frac{\zeta_k}{\sum_{l=0}^{n} \zeta_l}$$

Then the expected incremental losses satisfy

$$E[Z_{i,k}] = \alpha_i\,\vartheta_k$$

Therefore, the credibility model of Mack is a *multiplicative model* and the parameters $\vartheta_0, \vartheta_1, \ldots, \vartheta_n$ form a *development pattern for incremental quotas*.

Because of assumption (i) it is sufficient to consider, for any $i \in \{0, 1, \ldots, n\}$ and for any $s \in \{0, 1, \ldots, n\}$ such that $i + s \geq n + 1$, the credibility model

$$\{Z_{i,0}, Z_{i,1}, \ldots, Z_{i,n-i}, Z_{i,s}\}$$

Then each of these models is a credibility model with risk parameter Λ_i. The assumptions imply that

$$E[Z_{i,k}] = E[m_i(\Lambda_i)]\,\zeta_k$$
$$\mathrm{cov}[Z_{i,k}, Z_{i,l}] = E[v_{i,k}(\Lambda_i)]\,\delta_{k,l} + \mathrm{var}[m_i(\Lambda_i)]\,\zeta_k\,\zeta_l$$

Putting $X_{i,k} := Z_{i,k}/\zeta_k$ then yields

$$E[X_{i,k}] = E[m_i(\Lambda_i)]$$
$$\mathrm{cov}[X_{i,k}, X_{i,l}] = E[v_{i,k}(\Lambda_i)]\,\zeta_k^{-2}\,\delta_{k,l} + \mathrm{var}[m_i(\Lambda_i)]$$

Therefore, the family $\{X_{i,0}, X_{i,1}, \ldots, X_{i,n-i}, X_{i,s}\}$ of the scaled incremental losses fulfills the assumptions of the standard credibility model and with the definitions $\mu_i := E[m_i(\Lambda_i)]$ and $\tau_{i,k} := \zeta_k^2\,\mathrm{var}[m_i(\Lambda_i)]/E[v_{i,k}(\Lambda_i)]$ we obtain

$$X_{i,s}^{\mathrm{CR}} = \frac{1}{1 + \sum_{l=0}^{n-i} \tau_{i,l}}\,\mu_i + \sum_{k=0}^{n-i} \frac{\tau_{i,k}}{1 + \sum_{l=0}^{n-i} \tau_{i,l}}\,X_{i,k}$$

This yields

$$Z_{i,s}^{\mathrm{CR}} = \vartheta_s \left(\frac{1}{1 + \sum_{l=0}^{n-i} \tau_{i,l}}\,\mu_i + \sum_{k=0}^{n-i} \frac{\tau_{i,k}}{1 + \sum_{l=0}^{n-i} \tau_{i,l}}\,\frac{Z_{i,k}}{\vartheta_k} \right)$$

It is remarkable that the credibility predictors of all non-observable scaled incremental losses from the same accident year are identical.

The credibility reserves are then obtained by summation of the credibility predictors of the corresponding incremental losses.

Credibility Model of Witting

The model of Witting is a model for *claim numbers* and consists of the following assumptions:

Credibility Model of Witting:
  (i) *The incremental losses take only values in the set $\mathbb{N}_0$.*
 (ii) *Incremental losses from different accident years are uncorrelated.*
(iii) *The expected ultimate loss of every accident year $i \in \{0, 1, \ldots, n\}$ satisfies $E[S_{i,n}] > 0$.*
 (iv) *There exist parameters $\vartheta_0, \vartheta_1, \ldots, \vartheta_n \in (0, 1)$ with $\sum_{l=0}^{n} \vartheta_l = 1$ such that*

$$P_{Z_{i,0}, Z_{i,1}, \ldots, Z_{i,n}|S_{i,n}} = \mathbf{M}(S_{i,n}; \vartheta_0, \vartheta_1, \ldots, \vartheta_n)$$

holds for all $i \in \{0, 1, \ldots, n\}$.

The model of Witting is a *multinomial model* and hence a *multiplicative model* and the parameters $\vartheta_0, \vartheta_1, \ldots, \vartheta_n$ form a *development pattern for incremental quotas*.

We assume in this section that the assumptions of the model of Witting are fulfilled.

Because of assumption (ii) it is sufficient to consider, for any $i \in \{0, 1, \ldots, n\}$ and for any $s \in \{0, 1, \ldots, n\}$ such that $i + s \geq n + 1$, the credibility model

$$\{Z_{i,0}, Z_{i,1}, \ldots, Z_{i,n-i}, Z_{i,s}\}$$

Then each of these models is a credibility model with risk parameter $S_{i,n}$. To simplify the notation we define

$$\alpha_i := E[S_{i,n}]$$
$$\sigma_i^2 := \mathrm{var}[S_{i,n}]$$

Then the assumptions yield

$$E(Z_{i,k}|S_{i,n}) = S_{i,n}\,\vartheta_k,$$
$$\mathrm{cov}(Z_{i,k}, Z_{i,l}|S_{i,n}) = S_{i,n}\,\vartheta_k\,\delta_{k,l} - S_{i,n}\,\vartheta_k\vartheta_l$$

and hence

$$E[Z_{i,k}] = \alpha_i\,\vartheta_k$$
$$\mathrm{cov}[Z_{i,k}, Z_{i,l}] = \alpha_i\,\vartheta_k\,\delta_{k,l} + (\sigma_i^2 - \alpha_i)\,\vartheta_k\vartheta_l$$

Putting $X_{i,k} := Z_{i,k}/\vartheta_k$ then yields

$$E[X_{i,k}] = \alpha_i$$
$$\mathrm{cov}[X_{i,k}, X_{i,l}] = \alpha_i\,\vartheta_k^{-1}\,\delta_{k,l} + (\sigma_i^2 - \alpha_i)$$

Therefore, the family $\{X_{i,0}, X_{i,1}, \ldots, X_{i,n-i}, X_{i,s}\}$ of the scaled incremental losses fulfills the assumptions of the standard credibility model and with $\tau_i := (\sigma_i^2 - \alpha_i)/\alpha_i$ we obtain

$$X_{i,s}^{\mathrm{CR}} = \frac{1}{1 + \tau_i \sum_{l=0}^{n-i} \vartheta_l}\,\alpha_i + \sum_{k=0}^{n-i} \frac{\tau_i\,\vartheta_k}{1 + \tau_i \sum_{l=0}^{n-i} \vartheta_l}\,X_{i,k}$$

Letting $\gamma_{n-i} := \sum_{l=0}^{n-i} \vartheta_l$ this yields

$$Z_{i,s}^{\mathrm{CR}} = \vartheta_s \left(\frac{1}{1 + \tau_i\,\gamma_{n-i}}\,\alpha_i + \frac{\tau_i\,\gamma_{n-i}}{1 + \tau_i\,\gamma_{n-i}}\,\frac{S_{i,n-i}}{\gamma_{n-i}} \right)$$

Therefore, the credibility predictor $Z_{i,s}^{\text{CR}}$ depends on the observable incremental losses only via the current loss $S_{i,n-i}$. Furthermore, the sign of the parameter τ_i determines the manner in which the current loss contributes to the credibility predictor:

- In the case $\tau_i < 0$, the credibility predictor increases when the current loss increases; this case occurs when the ultimate loss has a *binomial distribution*.
- In the case $\tau_i = 0$, the credibility predictor does not depend on the current loss; this case occurs when the ultimate loss has a *Poisson distribution*.
- In the case $\tau_i > 0$, the credibility predictor increases when the current loss increases; this case occurs when the ultimate loss has a *negative binomial distribution*.

Note that $1 + \tau_i \, \gamma_{n-i} > 0$ holds in all cases.

Credibility Model of Hesselager and Witting

The model of Hesselager and Witting is a variation of the model of Witting. In this model, the fixed probabilities $\vartheta_0, \vartheta_1, \dots, \vartheta_n$ are replaced by random probabilities

$$\Theta_{i,0}, \Theta_{i,1}, \dots, \Theta_{i,n}$$

which depend on the accident year i but are independent of the ultimate loss $S_{i,n}$ of that accident year:

Credibility Model of Hesselager and Witting (special case):

(i) *The incremental losses take only values in the set $\mathbb{N}_0$.*

(ii) *Incremental losses from different accident years are uncorrelated.*

(iii) *The expected ultimate loss of every accident year $i \in \{0, 1, \dots, n\}$ satisfies $E[S_{i,n}] > 0$.*

(iv) *There exist parameters $\eta_0, \eta_1, \dots, \eta_n$ and a family of random variables $\{\Theta_{i,k}\}_{i,k \in \{0,1,\dots,n\}}$ such that, for every accident year $i \in \{0, 1, \dots, n\}$, the family $\{\Theta_{i,k}\}_{k \in \{0,1,\dots,n\}}$ is independent of $S_{i,n}$ and satisfies*

$$P_{\Theta_{i,0}, \Theta_{i,1}, \dots, \Theta_{i,n}} = \mathbf{Dir}(\eta_0, \eta_1, \dots, \eta_n)$$

and

$$P_{Z_{i,0}, Z_{i,1}, \dots, Z_{i,n} \mid S_{i,n}, \Theta_{i,0}, \Theta_{i,1}, \dots, \Theta_{i,n}} = \mathbf{M}(S_{i,n}; \Theta_{i,0}, \Theta_{i,1}, \dots, \Theta_{i,n})$$

The model of Hesselager and Witting is a *multiplicative model* and the expectations $E[\Theta_{i,0}], E[\Theta_{i,1}], \dots, E[\Theta_{i,n}]$, which by assumption are identical for all accident years, form a *development pattern for incremental quotas*. This is not immediately apparent but will become evident from the subsequent discussion of the model.

We assume in this section that the assumptions of the model of Hesselager and Witting are fulfilled.

Because of assumption (ii) it is sufficient to consider, for any $i \in \{0, 1, \dots, n\}$ and for any $s \in \{0, 1, \dots, n\}$ such that $i + s \geq n + 1$, the credibility model

$$\{Z_{i,0}, Z_{i,1}, \ldots, Z_{i,n-i}, Z_{i,s}\}$$

Then each of these models is a credibility model with a risk parameter Λ_i whose coordinates are the random variables $S_{i,n}, \Theta_{i,0}, \Theta_{i,1}, \ldots, \Theta_{i,n}$. To simplify the notation we define

$$\alpha_i := E[S_{i,n}]$$
$$\sigma_i^2 := \mathrm{var}[S_{i,n}]$$

Then the assumptions yield

$$E(Z_{i,k}|S_{i,n}, \Theta_{i,0}, \Theta_{i,1}, \ldots, \Theta_{i,n}) = S_{i,n}\,\Theta_{i,k}$$
$$\mathrm{cov}(Z_{i,k}, Z_{i,l}|S_{i,n}, \Theta_{i,0}, \Theta_{i,1}, \ldots, \Theta_{i,n}) = S_{i,n}\,\Theta_{i,k}\,\delta_{k,l} - S_{i,n}\,\Theta_{i,k}\Theta_{i,l}$$

and hence

$$E(Z_{i,k}|\Theta_{i,0}, \Theta_{i,1}, \ldots, \Theta_{i,n}) = \alpha_i\,\Theta_{i,k}$$
$$\mathrm{cov}(Z_{i,k}, Z_{i,l}|\Theta_{i,0}, \Theta_{i,1}, \ldots, \Theta_{i,n}) = \alpha_i\,\Theta_{i,k}\,\delta_{k,l} + (\sigma_i^2 - \alpha_i)\,\Theta_{i,k}\Theta_{i,l}$$

Putting $\eta := \sum_{l=0}^n \eta_l$ and $\vartheta_k := \eta_k/\eta$, the assumptions also yield

$$E[\Theta_{i,k}] = \vartheta_k$$
$$E[\Theta_{i,k}\Theta_{i,l}] = \frac{1}{1+\eta}\,\vartheta_k\,\delta_{k,l} + \frac{\eta}{1+\eta}\,\vartheta_k\vartheta_l$$

We thus obtain

$$E[Z_{i,k}] = \alpha_i\,\vartheta_k$$
$$\mathrm{cov}[Z_{i,k}, Z_{i,l}] = \frac{\sigma_i^2 + \alpha_i^2 + \eta\alpha_i}{1+\eta}\,\vartheta_k\,\delta_{k,l} + \frac{\eta\sigma_i^2 - \alpha_i^2 - \eta\alpha_i}{1+\eta}\,\vartheta_k\vartheta_l$$

Putting $X_{i,k} := Z_{i,k}/\vartheta_k$ then yields

$$E[X_{i,k}] = \alpha_i$$
$$\mathrm{cov}[X_{i,k}, X_{i,l}] = \frac{\sigma_i^2 + \alpha_i^2 + \eta\alpha_i}{1+\eta}\,\frac{1}{\vartheta_k}\,\delta_{k,l} + \frac{\eta\sigma_i^2 - \alpha_i^2 - \eta\alpha_i}{1+\eta}$$

Therefore, the family $\{X_{i,0}, X_{i,1}, \ldots, X_{i,n-i}, X_{i,s}\}$ of the scaled incremental losses fulfills the assumptions of the standard credibility model and with $\tau_i := (\eta\sigma_i^2 - \alpha_i^2 - \eta\alpha_i)/(\sigma_i^2 + \alpha_i^2 + \eta\alpha_i)$ we obtain

$$X_{i,s}^{\mathrm{CR}} = \frac{1}{1 + \tau_i \sum_{l=0}^{n-i} \vartheta_l}\,\alpha_i + \sum_{k=0}^{n-i} \frac{\tau_i\,\vartheta_k}{1 + \tau_i \sum_{l=0}^{n-i} \vartheta_l}\,X_{i,k}$$

Letting $\gamma_{n-i} := \sum_{l=0}^{n-i} \vartheta_l$ this yields

$$Z_{i,s}^{\mathrm{CR}} = \vartheta_s \left(\frac{1}{1 + \tau_i \, \gamma_{n-i}} \, \alpha_i + \frac{\tau_i \, \gamma_{n-i}}{1 + \tau_i \, \gamma_{n-i}} \, \frac{S_{i,n-i}}{\gamma_{n-i}} \right)$$

This identity has the same form as that obtained in the model of Witting and the concluding remark on the model of Witting also applies to the model of Hesselager and Witting.

Notes

Keywords: Credibility Models (Basics), Development Patterns (Basics), Multinomial Model, Multiplicative Models.

References: De Vylder [1982], Hess & Schmidt [2001], Hesselager & Witting [1988], Mack [1990, 2002], Schmidt [2012], Witting [1987].

Development Patterns (Basics)

Klaus D. Schmidt

Most models and methods of loss reserving are based on the assumptions that every claim is finally settled in either the accident year or one of n subsequent development years and that the development of the losses of a given accident year over the $n + 1$ development years follows a *development pattern*, which is identical for all accident years.

There are different possibilities to define a development pattern, but they are all based on the assumption of approximate proportionality of the accident years. In this article, we start with the development patterns for incremental quotas, (cumulative) quotas and factors, which turn out to be equivalent. We then consider the development pattern of Panning, which is less popular but completes the group of equivalent development patterns. Finally, we consider the development pattern for incremental loss ratios, which involves known volume measures of the accident years.

K.D. Schmidt (✉)
Technische Universität Dresden, Dresden, Germany
e-mail: klaus.d.schmidt@tu-dresden.de

© Springer International Publishing Switzerland 2016
M. Radtke et al. (eds.), *Handbook on Loss Reserving*,
EAA Series, DOI 10.1007/978-3-319-30056-6_13

Development Pattern for Incremental Quotas

Consider the run-off square of incremental losses:

Accident year	Development year					
	0	1	... k	... $n-i$	... $n-1$	n
0	$Z_{0,0}$	$Z_{0,1}$	... $Z_{0,k}$	... $Z_{0,n-i}$	... $Z_{0,n-1}$	$Z_{0,n}$
1	$Z_{1,0}$	$Z_{1,1}$	... $Z_{1,k}$	... $Z_{1,n-i}$	... $Z_{1,n-1}$	$Z_{1,n}$
$\vdots$	$\vdots$	$\vdots$	$\vdots$	$\vdots$	$\vdots$	$\vdots$
i	$Z_{i,0}$	$Z_{i,1}$	... $Z_{i,k}$	... $Z_{i,n-i}$	... $Z_{i,n-1}$	$Z_{i,n}$
$\vdots$	$\vdots$	$\vdots$	$\vdots$	$\vdots$	$\vdots$	$\vdots$
$n-k$	$Z_{n-k,0}$	$Z_{n-k,1}$	... $Z_{n-k,k}$	... $Z_{n-k,n-i}$	... $Z_{n-k,n-1}$	$Z_{n-k,n}$
$\vdots$	$\vdots$	$\vdots$	$\vdots$	$\vdots$	$\vdots$	$\vdots$
$n-1$	$Z_{n-1,0}$	$Z_{n-1,1}$	... $Z_{n-1,k}$	... $Z_{n-1,n-i}$	... $Z_{n-1,n-1}$	$Z_{n-1,n}$
n	$Z_{n,0}$	$Z_{n,1}$	... $Z_{n,k}$	... $Z_{n,n-i}$	... $Z_{n,n-1}$	$Z_{n,n}$

Let

$$S_{i,k} := \sum_{l=0}^{k} Z_{i,l}$$

be the cumulative loss from accident year i in development year k with $i, k \in \{0, 1, \ldots, n\}$.

The development pattern for incremental quotas compares expected incremental losses with the expected ultimate loss of the same accident year:

Development Pattern for Incremental Quotas: *There exist parameters* $\vartheta_0, \vartheta_1, \ldots, \vartheta_n$ *with* $\sum_{l=0}^{n} \vartheta_l = 1$ *such that the identity*

$$\frac{E[Z_{i,k}]}{E[S_{i,n}]} = \vartheta_k$$

holds for all $k \in \{0, 1, \ldots, n\}$ *and for all* $i \in \{0, 1, \ldots, n\}$.

In the development pattern for incremental quotas, the *incremental quota* ϑ_k depends on the development year k but not on the accident year i.

Development Pattern for (Cumulative) Quotas

Consider the run-off square for cumulative losses:

Accident year	Development year						
	0	1	... k	... $n-i$	... $n-1$	n	
0	$S_{0,0}$	$S_{0,1}$	... $S_{0,k}$	... $S_{0,n-i}$	... $S_{0,n-1}$	$S_{0,n}$	
1	$S_{1,0}$	$S_{1,1}$	... $S_{1,k}$	... $S_{1,n-i}$	... $S_{1,n-1}$	$S_{1,n}$	
$\vdots$	$\vdots$	$\vdots$	$\vdots$	$\vdots$	$\vdots$	$\vdots$	
i	$S_{i,0}$	$S_{i,1}$	... $S_{i,k}$	... $S_{i,n-i}$	... $S_{i,n-1}$	$S_{i,n}$	
$\vdots$	$\vdots$	$\vdots$	$\vdots$	$\vdots$	$\vdots$	$\vdots$	
$n-k$	$S_{n-k,0}$	$S_{n-k,1}$	... $S_{n-k,k}$	... $S_{n-k,n-i}$	... $S_{n-k,n-1}$	$S_{n-k,n}$	
$\vdots$	$\vdots$	$\vdots$	$\vdots$	$\vdots$	$\vdots$	$\vdots$	
$n-1$	$S_{n-1,0}$	$S_{n-1,1}$	... $S_{n-1,k}$	... $S_{n-1,n-i}$	... $S_{n-1,n-1}$	$S_{n-1,n}$	
n	$S_{n,0}$	$S_{n,1}$	... $S_{n,k}$	... $S_{n,n-i}$	... $S_{n,n-1}$	$S_{n,n}$	

The development pattern for (cumulative) quotas compares expected cumulative losses with the expected ultimate loss of the same accident year:

Development Pattern for (Cumulative) Quotas: *There exist parameters* $\gamma_0, \gamma_1, \ldots, \gamma_n$ *with* $\gamma_n = 1$ *such that the identity*

$$\frac{E[S_{i,k}]}{E[S_{i,n}]} = \gamma_k$$

holds for all $k \in \{0, 1, \ldots, n\}$ *and for all* $i \in \{0, 1, \ldots, n\}$.

In the development pattern for (cumulative) quotas, the (*cumulative*) *quota* γ_k of the expected cumulative loss $E[S_{i,k}]$ with respect to the expected ultimate loss $E[S_{i,n}]$ depends on the development year k but not on the accident year i.

Development Pattern for Factors

The development pattern for factors compares consecutive expected cumulative losses:

Development Pattern for Factors: *There exist parameters* $\varphi_1, \ldots, \varphi_n$ *such that the identity*

$$\frac{E[S_{i,k}]}{E[S_{i,k-1}]} = \varphi_k$$

holds for all $k \in \{1, \ldots, n\}$ *and for all* $i \in \{0, 1, \ldots, n\}$.

In the development pattern for factors, the *development factor* φ_k as the ratio between the expected cumulative loss $E[S_{i,k}]$ and the expected cumulative loss

$E[S_{i,k-1}]$ of the preceding development year depends on the development year k but not on the accident year i.

Comparison

The following theorem shows that the development patterns for incremental quotas, quotas and factors are equivalent:

Theorem.

(1) Let $\vartheta_0, \vartheta_1, \ldots, \vartheta_n$ be a development pattern for incremental quotas and define

$$\gamma_k := \sum_{l=0}^{k} \vartheta_l$$

for all $k \in \{0, 1, \ldots, n\}$. Then $\gamma_0, \gamma_1, \ldots, \gamma_n$ is a development pattern for quotas.

(2) Let $\gamma_0, \gamma_1, \ldots, \gamma_n$ be a development pattern for quotas and define

$$\varphi_k := \gamma_k / \gamma_{k-1}$$

for all $k \in \{1, \ldots, n\}$. Then $\varphi_1, \ldots, \varphi_n$ is a development pattern for factors.

(3) Let $\varphi_1, \ldots, \varphi_n$ be a development pattern for factors and define

$$\gamma_k := \prod_{l=k+1}^{n} \frac{1}{\varphi_l}$$

for all $k \in \{0, 1, \ldots, n\}$. Then $\gamma_0, \gamma_1, \ldots, \gamma_n$ is a development pattern for quotas.

(4) Let $\gamma_0, \gamma_1, \ldots, \gamma_n$ be a development pattern for quotas and define

$$\vartheta_k := \begin{cases} \gamma_0 & \text{if } k = 0 \\ \gamma_k - \gamma_{k-1} & \text{else} \end{cases}$$

for all $k \in \{0, 1, \ldots, n\}$. Then $\vartheta_0, \vartheta_1, \ldots, \vartheta_n$ is a development pattern for incremental quotas.

As a corollary of the theorem we obtain a relation between the development pattern for incremental quotas and the development pattern for factors:

Corollary.

(1) Let $\vartheta_0, \vartheta_1, \ldots, \vartheta_n$ be a development pattern for incremental quotas and define

$$\varphi_k := \frac{\sum_{l=0}^{k} \vartheta_l}{\sum_{l=0}^{k-1} \vartheta_l}$$

for all $k \in \{1, \ldots, n\}$. Then $\varphi_1, \ldots, \varphi_n$ is a development pattern for factors.

(2) *Let $\varphi_1, \ldots, \varphi_n$ be a development pattern for factors and define*

$$
\vartheta_k := \begin{cases} \prod_{l=1}^{n} \dfrac{1}{\varphi_l} & \text{if } k = 0 \\[2ex] \prod_{l=k+1}^{n} \dfrac{1}{\varphi_l} - \prod_{l=k}^{n} \dfrac{1}{\varphi_l} & \text{else} \end{cases}
$$

for all $k \in \{0, 1, \ldots, n\}$. Then $\vartheta_0, \vartheta_1, \ldots, \vartheta_n$ is a development pattern for incremental quotas.

The assumption that one, and hence each, of the development patterns for incremental quotas, quotas or factors exists is in fact an assumption on the joint distribution of all incremental losses or cumulative losses and hence an assumption on the existence of a particular stochastic model. In all three cases, the stochastic model is as simple as possible since its assumptions only involve expected values.

Example. The three rows of the following table contain

- a development pattern for incremental quotas,
- a development pattern for quotas, and
- a development pattern for factors.

Each of these development patterns determines the two other development patterns. Note that, by definition, the quota for the last development year is equal to 1 and there is no development factor for development year 0.

	Development year k			
	0	1	2	3
ϑ_k	0.400	0.240	0.160	0.200
γ_k	0.400	0.640	0.800	1
φ_k		1.600	1.250	1.250

In practice the development patterns are unknown and have to be estimated.

Depending on the type of the *run-off data* underlying the run-off triangle, the three development patterns may possess additional properties:

Theorem. *Assume that*

- *$\vartheta_0, \vartheta_1, \ldots, \vartheta_n$ is a development pattern for incremental quotas,*
- *$\gamma_0, \gamma_1, \ldots, \gamma_n$ is a development pattern for quotas and*
- *$\varphi_1, \ldots, \varphi_n$ is a development pattern for factors*

such that these three development patterns result from each other. The following properties are equivalent:

(a) *The inequality $\vartheta_k > 0$ holds for all $k \in \{0, 1, \ldots, n\}$.*
(b) *The quotas satisfy $0 < \gamma_0 < \gamma_1 < \cdots < \gamma_n$.*
(c) *The inequality $\varphi_k > 1$ holds for all $k \in \{1, \ldots, n\}$.*

Usually, the equivalent conditions of the theorem are fulfilled when the run-off data consists of *numbers of claims* or *paid losses*, but they are not fulfilled when the run-off data consists of *incurred losses*.

To complete the discussion of these three development patterns, let us note that a development pattern for incremental quotas, quotas or factors exists if and only if there exists a *multiplicative model*.

Development Pattern of Panning

The development pattern of Panning compares the expected incremental losses with the expected incremental losses of development year 0:

Development Pattern of Panning: *There exist parameters $\xi_0, \xi_1, \ldots, \xi_n$ with $\xi_0 = 1$ such that the identity*

$$\frac{E[Z_{i,k}]}{E[Z_{i,0}]} = \xi_k$$

holds for all $k \in \{0, 1, \ldots, n\}$ and for all $i \in \{0, 1, \ldots, n\}$.

In the development pattern of Panning the ratio ξ_k between the expected incremental losses $E[Z_{i,k}]$ and $E[Z_{i,0}]$ depends on the development year k but not on the accident year i. The development pattern of Panning is equivalent to the development patterns for incremental quotas, quotas and factors:

Theorem.

(1) *Let $\xi_0, \xi_1, \ldots, \xi_n$ be a development pattern of Panning and define*

$$\vartheta_k := \frac{\xi_k}{\sum_{l=0}^{n} \xi_l}$$

for all $k \in \{0, 1, \ldots, n\}$. Then $\vartheta_0, \vartheta_1, \ldots, \vartheta_n$ is a development pattern for incremental quotas.

(2) *Let $\vartheta_0, \vartheta_1, \ldots, \vartheta_n$ be a development pattern for incremental quotas and define*

$$\xi_k := \vartheta_k / \vartheta_0$$

for all $k \in \{0, 1, \ldots, n\}$. Then $\xi_0, \xi_1, \ldots, \xi_n$ is a development pattern of Panning.

Example. The rows of the following table contain

- a development pattern of Panning and
- a development pattern for incremental quotas.

Each of these development patterns determines the other development pattern.

	Development year k			
	0	1	2	3
ξ_k	1	0.600	0.400	0.500
ϑ_k	0.400	0.240	0.160	0.200

In practice the development patterns are unknown and have to be estimated.

Development Pattern for Incremental Loss Ratios

The development pattern for incremental loss ratios refers to known *volume measures* $v_0, v_1, \ldots, v_n$ of the accident years. It compares the expected incremental losses with the volume measures:

> **Development Pattern for Incremental Loss Ratios**: *There exist parameters* $\zeta_0, \zeta_1, \ldots, \zeta_n$ *such that the identity*
>
> $$E\left[\frac{Z_{i,k}}{v_i}\right] = \frac{E[Z_{i,k}]}{v_i} = \zeta_k$$
>
> *holds for all* $k \in \{0, 1, \ldots, n\}$ *and for all* $i \in \{0, 1, \ldots, n\}$.

In the development pattern for incremental loss ratios, the *expected incremental loss ratio* ζ_k depends on the development year k but not on the accident year i. A connection between the development pattern for incremental loss ratios and the development pattern for incremental quotas can be established by means of the *expected ultimate loss ratios* $E[S_{i,n}/v_i]$:

Theorem.

(1) *Let* $\zeta_0, \zeta_1, \ldots, \zeta_n$ *be a development pattern for incremental loss ratios and let*

$$\kappa := \sum_{l=0}^{n} \zeta_l$$

Define
$$\vartheta_k := \zeta_k/\kappa$$

for all $k \in \{0, 1, \ldots, n\}$. *Then* $\vartheta_0, \vartheta_1, \ldots, \vartheta_n$ *is a development pattern for incremental quotas and the parameter* κ *satisfies*

$$E\left[\frac{S_{i,n}}{v_i}\right] = \kappa$$

for all $i \in \{0, 1, \ldots, n\}$.

(2) *Let* $\vartheta_0, \vartheta_1, \ldots, \vartheta_n$ *be a development pattern for incremental quotas and assume that there exists some parameter* κ *such that the identity*

$$E\left[\frac{S_{i,n}}{v_i}\right] = \kappa$$

holds for all $i \in \{0, 1, \ldots, n\}$. *Define*

$$\zeta_k := \kappa\,\vartheta_k$$

for all $k \in \{0, 1, \ldots, n\}$. *Then* $\zeta_0, \zeta_1, \ldots, \zeta_n$ *is a development pattern for incremental loss ratios.*

Therefore, a development pattern for incremental loss ratios exists if and only if there exists a development pattern for incremental quotas (or quotas or factors) and the expected ultimate loss ratios are equal for all accident years.

Example. The rows of the following table contain

- a development pattern for incremental loss ratios as well as the ultimate loss ratio, which is equal for all accident years, and
- a development pattern for incremental quotas.

The development pattern for incremental loss ratios is equivalent to the development pattern for incremental quotas together with the ultimate loss ratio.

	Development year k				κ
	0	1	2	3	
ζ_k	0.320	0.192	0.128	0.160	0.800
ϑ_k	0.400	0.240	0.160	0.200	

In practice the development patterns and the expected ultimate loss ratio are unknown and have to be estimated.

Other Development Patterns

The development patterns for incremental quotas, quotas or factors and the development pattern of Panning use the assumption that the quotients of certain expected values are identical for all accident years. This does not completely meet intuition, which would suggest considering expectations of quotients rather than quotients of expectations.

In the case of the development pattern for incremental quotas this means that the assumption

$$E\left[\frac{Z_{i,k}}{S_{i,n}}\right] = \vartheta_k$$

for all $i \in \{0, 1, \ldots, n\}$ would be preferred to the assumption

$$\frac{E[Z_{i,k}]}{E[S_{i,n}]} = \vartheta_k$$

for all $i \in \{0, 1, \ldots, n\}$. These assumptions are equivalent if and only if

$$\mathrm{cov}\left[\frac{Z_{i,k}}{S_{i,n}}, S_{i,n}\right] = 0$$

holds for all $i \in \{0, 1, \ldots, n\}$, which means that the incremental quotas $Z_{i,k}/S_{i,n}$ and the ultimate losses $S_{i,n}$ are uncorrelated.

It turns out that the definitions of the development patterns given before are reasonable since they provide a possibility for the justification and the comparison of several methods of loss reserving.

The development patterns considered here can be refined. For example, the assumption that there exist parameters $\vartheta_0, \vartheta_1, \ldots, \vartheta_n$ such that the identity

$$E\left(\frac{Z_{i,k}}{S_{i,n}} \,\middle|\, S_{i,n}\right) = \vartheta_k$$

holds for all $k \in \{0, 1, \ldots, n\}$ and for all $i \in \{0, 1, \ldots, n\}$, implies

$$E\left[\frac{Z_{i,k}}{S_{i,n}}\right] = \vartheta_k = \frac{E[Z_{i,k}]}{E[S_{i,n}]}$$

and hence

$$\mathrm{cov}\left[\frac{Z_{i,k}}{S_{i,n}}, S_{i,n}\right] = 0$$

The assumption of such a *conditional development pattern for incremental quotas* is fulfilled in the *multinomial model* and in particular in the *Poisson model*. Another conditional development pattern is used in the *chain ladder models* of Mack and Schnaus.

Infinite Development Pattern and Tail

The development patterns considered so far rely on the assumption that there exists a fixed *development horizon* n in the sense that all claims are settled after n years. This assumption is not realistic.

Formally, the problem can be solved by the assumption that there is an infinite number of development years and by replacing for instance the development pattern $\varphi_1, \ldots, \varphi_n$ for factors with an *infinite development pattern* $\{\varphi_k\}_{k \in \mathbb{N}}$. As infinite development patterns are difficult to handle, a compromise can be made by introducing a *tail* of only d additional development years and replacing the development pattern $\varphi_1, \ldots, \varphi_n$ with an *extended development pattern* $\varphi_1, \ldots, \varphi_n, \varphi_{n+1}, \ldots, \varphi_{n+d}$. The incremental and cumulative losses of the tail are non-observable.

Does a Development Pattern Exist?

There is no objective answer to the question of whether or not the development of
losses of a given portfolio is subject to a development pattern. Instead, the answer
always consists in the decision of the actuary to accept or to refuse the assumption of
the existence of a development pattern. Such a decision requires a detailed analysis
of the portfolio and can be supported by the comparison of the observable individual
development of the single accident years. We illustrate this by an example:

Example. Consider the run-off triangle of incremental losses with known volume
measures of the accident years

Accident year i	\multicolumn{6}{c}{Development year k}	Volume v_i
	0	1	2	3	4	5	
0	1001	854	568	565	347	148	4025
1	1113	990	671	648	422		4456
2	1265	1168	800	744			5315
3	1490	1383	1007				5986
4	1725	2536					6939
5	1889						8158

and the corresponding run-off triangle of cumulative losses

Accident year i	\multicolumn{6}{c}{Development year k}	Volume v_i
	0	1	2	3	4	5	
0	1001	1855	2423	2988	3335	3483	4025
1	1113	2103	2774	3422	3844		4456
2	1265	2433	3233	3977			5315
3	1490	2873	3880				5986
4	1725	4261					6939
5	1889						8158

We obtain the following table for the observable *individual development factors*
$S_{i,k}/S_{i,k-1}$

Accident year i	\multicolumn{6}{c}{Development year k}	Volume v_i
	0	1	2	3	4	5	
0		1.853	1.306	1.233	1.116	1.044	4025
1		1.889	1.319	1.234	1.123		4456
2		1.923	1.329	1.230			5315
3		1.928	1.351				5986
4		2.470					6939
5							8158

and for the observable *individual incremental loss ratios* $Z_{i,k}/v_i$ we obtain

Accident	Development year k						Volume
year i	0	1	2	3	4	5	v_i
0	0.249	0.212	0.141	0.140	0.086	0.037	4025
1	0.250	0.222	0.151	0.145	0.095		4456
2	0.238	0.220	0.151	0.140			5315
3	0.249	0.231	0.168				5986
4	0.249	0.365					6939
5	0.232						8158

The tables show that the loss development in old accident years runs quite uniformly. By contrast, the values of $S_{4,1}/S_{4,0}$ and $Z_{4,1}/v_4$ are eye-catching and give reason to examine the cause for the outliers $Z_{4,1}$ and $S_{4,1}$.

The analysis of the portfolio may lead to the decision to accept the assumption that a development pattern exists; in this case, the actuary still has to decide whether the development pattern should be estimated

- based only on the run-off data of the portfolio or
- based on the run-off data of the portfolio combined with appropriate volume measures or
- based on data from comparable portfolios or
- based on market statistics.

On the other hand, the analysis of the portfolio may also lead to the insight that the development in younger accident years deviates substantially from that of older accident years, so that the assumption of a development pattern is not acceptable.

Notes

Keywords: Aggregation, Chain Ladder Method (Models), Development Patterns (Estimation), Loss Ratios, Multinomial Model, Multiplicative Models, Poisson Model, Run-Off Data, Run-Off Triangles, Tail Estimation.

References: GDV [2011], Reich & Zeller [1988], Schmidt [2009, 2012], Schmidt & Zocher [2008], Taylor [2000].

Development Patterns (Estimation)

Klaus D. Schmidt

In practice the parameters of a development pattern are unknown and have to be estimated. In this article we consider estimation of the parameters of a development pattern.

Development Patterns for Incremental Quotas and Quotas

The development patterns for incremental quotas and quotas use the expected ultimate losses. As only the ultimate loss of accident year 0 is observable, only the *individual incremental quota*

$$\widehat{\vartheta}_{0,k} := Z_{0,k}/S_{0,n}$$

and the *individual (cumulative) quota*

$$\widehat{\gamma}_{0,k} := S_{0,k}/S_{0,n}$$

of accident year 0 are available for the estimation of the incremental quota $\vartheta_k = E[Z_{i,k}]/E[S_{i,n}]$ or the (cumulative) quota $\gamma_k = E[S_{i,k}]/E[S_{i,n}]$, respectively. This means that only a minor part of the data given by the run-off triangle can be used for direct estimation of the parameters.

We show below that estimators of incremental quotas and quotas, which rely on a larger part of the run-off triangle, can be constructed by using the equivalence of these development patterns with other development patterns.

K.D. Schmidt (✉)
Technische Universität Dresden, Dresden, Germany
e-mail: klaus.d.schmidt@tu-dresden.de

© Springer International Publishing Switzerland 2016
M. Radtke et al. (eds.), *Handbook on Loss Reserving*,
EAA Series, DOI 10.1007/978-3-319-30056-6_14

119

Development Pattern for Factors

For the estimation of the development factor $\varphi_k = E[S_{i,k}]/E[S_{i,k-1}]$, all *individual development factors*

$$\widehat{\varphi}_{i,k} := S_{i,k}/S_{i,k-1}$$

with $i \in \{0, 1, \ldots, n-k\}$ are available, and this is also true for their weighted means

$$\widehat{\varphi}_k = \sum_{j=0}^{n-k} W_{j,k}\,\widehat{\varphi}_{j,k}$$

with (random or constant) weights $W_{j,k}$ with $\sum_{j=0}^{n-k} W_{j,k} = 1$. An example of such a weighted mean is the *chain ladder factor*

$$\varphi_k^{CL} := \frac{\sum_{j=0}^{n-k} S_{j,k}}{\sum_{j=0}^{n-k} S_{j,k-1}} = \sum_{j=0}^{n-k} \frac{S_{j,k-1}}{\sum_{h=0}^{n-k} S_{h,k-1}} \frac{S_{j,k}}{S_{j,k-1}} = \sum_{j=0}^{n-k} \frac{S_{j,k-1}}{\sum_{h=0}^{n-k} S_{h,k-1}}\,\widehat{\varphi}_{j,k}$$

which is used in the *chain ladder method*.

Using arbitrary estimators $\widehat{\varphi}_1, \ldots, \widehat{\varphi}_n$ of the development pattern $\varphi_1, \ldots, \varphi_n$ of factors and appropriate conversion formulas for development patterns, we obtain estimators

$$\widehat{\vartheta}_k := \begin{cases} \displaystyle\prod_{l=1}^{n} \frac{1}{\widehat{\varphi}_l} & \text{if } k = 0 \\[2em] \displaystyle\prod_{l=k+1}^{n} \frac{1}{\widehat{\varphi}_l} - \prod_{l=k}^{n} \frac{1}{\widehat{\varphi}_l} & \text{else} \end{cases}$$

of the development pattern for incremental quotas with $\sum_{k=0}^{n} \widehat{\vartheta}_k = 1$ and estimators

$$\widehat{\gamma}_k := \prod_{l=k+1}^{n} \frac{1}{\widehat{\varphi}_l}$$

of the development pattern for quotas with $\widehat{\gamma}_n = 1$.

Development Pattern of Panning

For the estimation of the parameter $\xi_k = E[Z_{i,k}]/E[Z_{i,0}]$ of the development pattern of Panning, all *individual incremental development factors*

$$\widehat{\xi}_{i,k} := Z_{i,k}/Z_{i,0}$$

with $i \in \{0, 1, \ldots, n-k\}$ are available, and this is also true for their weighted means

$$\widehat{\xi}_k = \sum_{j=0}^{n-k} W_{j,k} \widehat{\xi}_{j,k}$$

with (random or constant) weights $W_{j,k}$ with $\sum_{j=0}^{n-k} W_{j,k} = 1$. An example of such a weighted mean is the *Panning factor*

$$\xi_k^{PA} := \frac{\sum_{j=0}^{n-k} Z_{j,0} Z_{j,k}}{\sum_{j=0}^{n-k} Z_{j,0}^2} = \sum_{j=0}^{n-k} \frac{Z_{j,0}^2}{\sum_{h=0}^{n-k} Z_{h,0}^2} \frac{Z_{j,k}}{Z_{j,0}} = \sum_{j=0}^{n-k} \frac{Z_{j,0}^2}{\sum_{h=0}^{n-k} Z_{h,0}^2} \widehat{\xi}_{j,k}$$

which is used in the *Panning method*.

Using arbitrary estimators $\widehat{\xi}_0, \widehat{\xi}_1, \ldots, \widehat{\xi}_n$ of the development pattern $\xi_0, \xi_1, \ldots, \xi_n$ of Panning and appropriate conversion formulas for development patterns, we obtain estimators

$$\widehat{\vartheta}_k := \frac{\widehat{\xi}_k}{\sum_{l=0}^{n} \widehat{\xi}_l}$$

of the development pattern for incremental quotas with $\sum_{k=0}^{n} \widehat{\vartheta}_k = 1$ and estimators

$$\widehat{\gamma}_k := \frac{\sum_{l=0}^{k} \widehat{\xi}_l}{\sum_{l=0}^{n} \widehat{\xi}_l}$$

of the development pattern for quotas with $\widehat{\gamma}_n = 1$.

Development Pattern for Incremental Loss Ratios

For the estimation of the expected incremental loss ratio $\zeta_k = E[Z_{i,k}/v_i]$, all *individual incremental loss ratios*

$$\widehat{\zeta}_{i,k} := Z_{i,k}/v_i$$

with $i \in \{0, 1, \ldots, n-k\}$ are available, and this is also true for their weighted means

$$\widehat{\zeta}_k = \sum_{j=0}^{n-k} W_{j,k} \widehat{\zeta}_{j,k}$$

with (random or constant) weights $W_{j,k}$ with $\sum_{j=0}^{n-k} W_{j,k} = 1$. An example of such a weighted mean is the *additive incremental loss ratio*

$$\zeta_k^{\mathrm{AD}} := \frac{\sum_{j=0}^{n-k} Z_{j,k}}{\sum_{j=0}^{n-k} v_j} = \sum_{j=0}^{n-k} \frac{v_j}{\sum_{h=0}^{n-k} v_h} \frac{Z_{j,k}}{v_j} = \sum_{j=0}^{n-k} \frac{v_j}{\sum_{h=0}^{n-k} v_h} \widehat{\zeta}_{j,k}$$

which is used in the *additive method*.

Using arbitrary estimators $\widehat{\zeta}_0, \widehat{\zeta}_1, \ldots, \widehat{\zeta}_n$ of the development pattern $\zeta_0, \zeta_1, \ldots, \zeta_n$ and appropriate conversion formulas for development patterns, we obtain estimators

$$\widehat{\vartheta}_k := \frac{\widehat{\zeta}_k}{\sum_{l=0}^{n} \widehat{\zeta}_l}$$

of the development pattern for incremental quotas with $\sum_{k=0}^{n} \widehat{\vartheta}_k = 1$ and estimators

$$\widehat{\gamma}_k := \frac{\sum_{l=0}^{k} \widehat{\zeta}_l}{\sum_{l=0}^{n} \widehat{\zeta}_l}$$

of the development pattern for quotas with $\widehat{\gamma}_n = 1$.

Remarks

In the case of changes of the structure of the portfolio or the development behaviour of the company, the data from old accident years or old calendar years may have very limited significance for the purpose of prediction. In such cases, the weights in the weighted means used as estimators of development patterns should be adjusted in an appropriate way.

In the case of scarce run-off data, which in particular may occur in *new business*, it can be reasonable to use estimators of development patterns from comparable portfolios or from market statistics.

Notes

Keywords: Additive Method, Chain Ladder Method (Basics), Development Patterns (Basics), Panning Method.

References: GDV [2011], Schmidt [2009, 2012], Schmidt & Zocher [2008].

Expected Loss Method

Anja Schnaus

Consider the run-off square of cumulative losses:

Accident year	Development year						
	0	1	... k	... $n-i$	... $n-1$	n	
0	$S_{0,0}$	$S_{0,1}$	... $S_{0,k}$	... $S_{0,n-i}$	... $S_{0,n-1}$	$S_{0,n}$	
1	$S_{1,0}$	$S_{1,1}$	... $S_{1,k}$	... $S_{1,n-i}$	... $S_{1,n-1}$	$S_{1,n}$	
$\vdots$	$\vdots$	$\vdots$	$\vdots$	$\vdots$	$\vdots$	$\vdots$	
i	$S_{i,0}$	$S_{i,1}$	... $S_{i,k}$	... $S_{i,n-i}$	... $S_{i,n-1}$	$S_{i,n}$	
$\vdots$	$\vdots$	$\vdots$	$\vdots$	$\vdots$	$\vdots$	$\vdots$	
$n-k$	$S_{n-k,0}$	$S_{n-k,1}$	... $S_{n-k,k}$	... $S_{n-k,n-i}$	... $S_{n-k,n-1}$	$S_{n-k,n}$	
$\vdots$	$\vdots$	$\vdots$	$\vdots$	$\vdots$	$\vdots$	$\vdots$	
$n-1$	$S_{n-1,0}$	$S_{n-1,1}$	... $S_{n-1,k}$	... $S_{n-1,n-i}$	... $S_{n-1,n-1}$	$S_{n-1,n}$	
n	$S_{n,0}$	$S_{n,1}$	... $S_{n,k}$	... $S_{n,n-i}$	... $S_{n,n-1}$	$S_{n,n}$	

We assume that the cumulative losses $S_{i,k}$ are observable for $i + k \leq n$ and that they are non-observable for $i + k \geq n + 1$.

The expected loss method is a very simple method of prediction, which is primarily suitable for *new business*. It uses only the current losses $S_{i,n-i}$ as well as *a priori estimators*

$$\widehat{\alpha}_0, \widehat{\alpha}_1, \ldots, \widehat{\alpha}_n$$

of the expected ultimate losses $\alpha_0, \alpha_1, \ldots, \alpha_n$. The a priori estimators are typically based on external information, which is not contained in the run-off triangle (for example on market statistics, on similar portfolios or on pricing assumptions).

A. Schnaus (✉)
Kerpen, Germany
e-mail: schnaus@genre.com

© Springer International Publishing Switzerland 2016
M. Radtke et al. (eds.), *Handbook on Loss Reserving*,
EAA Series, DOI 10.1007/978-3-319-30056-6_15

The expected loss method does not use the a priori estimators of the expected ultimate losses for the prediction of the future cumulative or incremental losses, but exclusively for the straight prediction of the accident year reserves and the aggregate loss reserve.

The *expected loss predictors* of the accident year reserves R_i for accident years $i \in \{1, \ldots, n\}$ are defined by

$$R_i^{\text{EL}} := \max\left\{\widehat{\alpha}_i - S_{i,n-i}, 0\right\}$$

and the *expected loss predictor* of the aggregate loss reserve R is defined by

$$R^{\text{EL}} := \sum_{j=1}^{n} R_j^{\text{EL}}$$

These predictors of the reserves are called *expected loss reserves*.

Example A. Calculation of the expected loss predictors of the accident year reserves and the aggregate loss reserve:

Accident	Development year k						Reserve
year i	0	1	2	3	4	5	$\widehat{\alpha}_i$	R_i^{EL}
0						3483		
1					3844		4200	356
2				3977			4800	823
3			3880				5400	1520
4		3261					6000	2739
5	1889						6600	4711
Sum								10149

Prediction of calender year reserves is not possible.

Example B. In this example the cumulative loss $S_{4,1}$ is increased by 1000:

Accident	Development year k						reserve
year i	0	1	2	3	4	5	$\widehat{\alpha}_i$	R_i^{EL}
0						3483		
1					3844		4200	356
2				3977			4800	823
3			3880				5400	1520
4		4261					6000	1739
5	1889						6600	4711
Sum								9149

The outlier $S_{4,1}$ is compensated by the expected loss predictors of the accident year reserve R_4 and of the aggregate loss reserve.

The expected loss method has numerous disadvantages:

- The a priori estimators of the expected ultimate losses can be generated without consideration of the observable cumulative losses of the run-off triangle.
- The expected loss accident year reserve is just the difference between the a priori estimator of the expected ultimate loss and the current loss. This implies that the accident year reserve is too low in case of present *large losses* and too high in case of *absent losses*, and this is also true for the expected loss aggregate loss reserve.
- The expected loss method does not use a *development pattern*. Due to the missing development pattern the future incremental losses cannot be predicted, and hence there is no possibility to break down the expected loss aggregate loss reserve to the individual calendar years in a reasonable way.

Because of this disadvantage the expected loss method is practically only appropriate for *new business*.

The *Bornhuetter–Ferguson method* uses not only the a priori estimators of the expected ultimate losses but also a development pattern and thus avoids, at least to some extent, the disadvantages of the expected loss method.

Remark

In practice the a priori estimators of the expected ultimate losses are often determined by means of *volume measures* and estimators of the corresponding *expected ultimate loss ratios*.

Notes

Keywords: Bornhuetter–Ferguson Method, Development Patterns (Basics), Loss Ratios, Run-Off Triangles.

Reference: Bornhuetter & Ferguson [1972].

Grossing up Method

Holger Lorenz and Klaus D. Schmidt

Consider the run-off square of cumulative losses:

Accident year	Development year						
	0	1	... k	... $n-i$	... $n-1$	n	
0	$S_{0,0}$	$S_{0,1}$	... $S_{0,k}$	... $S_{0,n-i}$	... $S_{0,n-1}$	$S_{0,n}$	
1	$S_{1,0}$	$S_{1,1}$	... $S_{1,k}$	... $S_{1,n-i}$	... $S_{1,n-1}$	$S_{1,n}$	
$\vdots$	$\vdots$	$\vdots$	$\vdots$	$\vdots$	$\vdots$	$\vdots$	
i	$S_{i,0}$	$S_{i,1}$	... $S_{i,k}$	... $S_{i,n-i}$	... $S_{i,n-1}$	$S_{i,n}$	
$\vdots$	$\vdots$	$\vdots$	$\vdots$	$\vdots$	$\vdots$	$\vdots$	
$n-k$	$S_{n-k,0}$	$S_{n-k,1}$	... $S_{n-k,k}$	... $S_{n-k,n-i}$	... $S_{n-k,n-1}$	$S_{n-k,n}$	
$\vdots$	$\vdots$	$\vdots$	$\vdots$	$\vdots$	$\vdots$	$\vdots$	
$n-1$	$S_{n-1,0}$	$S_{n-1,1}$	... $S_{n-1,k}$	... $S_{n-1,n-i}$	... $S_{n-1,n-1}$	$S_{n-1,n}$	
n	$S_{n,0}$	$S_{n,1}$	... $S_{n,k}$	... $S_{n,n-i}$	... $S_{n,n-1}$	$S_{n,n}$	

We assume that the cumulative losses $S_{i,k}$ are observable for $i + k \leq n$ and that they are non-observable for $i + k \geq n + 1$.

The grossing up method is a prediction method based on the development pattern for quotas:

Development Pattern for Quotas: *There exist parameters $\gamma_0, \gamma_1, \ldots, \gamma_n$ with $\gamma_n = 1$ such that the identity*

$$\frac{E[S_{i,k}]}{E[S_{i,n}]} = \gamma_k$$

holds for all $k \in \{0, 1, \ldots, n\}$ and for all $i \in \{0, 1, \ldots, n\}$.

H. Lorenz · K.D. Schmidt (✉)
Technische Universität Dresden, Dresden, Germany
e-mail: klaus.d.schmidt@tu-dresden.de

© Springer International Publishing Switzerland 2016
M. Radtke et al. (eds.), *Handbook on Loss Reserving*,
EAA Series, DOI 10.1007/978-3-319-30056-6_16

127

In this article we assume that a development pattern for quotas exists. Then the parameters $\vartheta_0, \vartheta_1, \ldots, \vartheta_n$ with

$$\vartheta_k := \begin{cases} \gamma_0 & \text{if } k = 0 \\ \gamma_k - \gamma_{k-1} & \text{else} \end{cases}$$

form a *development pattern for incremental quotas*. In particular, we are in the situation of a *multiplicative model*.

The *grossing up method* is defined by recursion:

- Define

$$\gamma_n^{GU} := 1$$

and

$$S_{0,n}^{GU} := S_{0,n}$$

- For every accident year $i \in \{1, \ldots, n\}$, the quota γ_{n-i} is estimated by the *grossing up quota*

$$\gamma_{n-i}^{GU} := \frac{\sum_{j=0}^{i-1} S_{j,n-i}}{\sum_{j=0}^{i-1} S_{j,n}^{GU}}$$

and the ultimate loss $S_{i,n}$ is predicted by the *grossing up predictor*

$$S_{i,n}^{GU} := \frac{S_{i,n-i}}{\gamma_{n-i}^{GU}}$$

More generally, for every accident year i and every development year k such that $i + k \geq n + 1$, the future cumulative loss $S_{i,k}$ is predicted by the *grossing up predictor*

$$S_{i,k}^{GU} := \gamma_k^{GU} S_{i,n}^{GU} = \gamma_k^{GU} \frac{S_{i,n-i}}{\gamma_{n-i}^{GU}}$$

The definition of the grossing up predictors of the future cumulative losses replicates the identity

$$E[S_{i,k}] = \frac{E[S_{i,k}]}{E[S_{i,n}]} \frac{E[S_{i,n}]}{E[S_{i,n-i}]} E[S_{i,n-i}] = \gamma_k \frac{1}{\gamma_{n-i}} E[S_{i,n-i}]$$

which results from the development pattern for quotas.

Example A. Calculation of the grossing up predictors of cumulative losses:

Accident year i	Development year k					
	0	1	2	3	4	5
0	1001	1855	2423	2988	3335	3483
1	1113	2103	2774	3422	3844	4015
2	1265	2433	3233	3977	4454	4652
3	1490	2873	3880	4781	5354	5592
4	1725	3261	4333	5339	5980	6242
5	1889	3588	4768	5875	6579	6871
γ_k^{GU}	0.27	0.52	0.69	0.85	0.96	1

This table is identical with that for the chain ladder method.

Example B. In this example the cumulative loss $S_{4,1}$ is increased by 1000:

Accident year i	Development year k					
	0	1	2	3	4	5
0	1001	1855	2423	2988	3335	3483
1	1113	2103	2774	3422	3844	4015
2	1265	2433	3233	3977	4454	4652
3	1490	2873	3880	4781	5354	5592
4	1725	4261	5662	6976	7813	8160
5	1889	3875	5148	6344	7105	7420
γ_k^{GU}	0.25	0.52	0.69	0.85	0.96	1

Again, this table is identical with that for the chain ladder method.

Comparison with the Chain Ladder Method

As the development patterns for quotas and factors are equivalent, the grossing up method is based on the same assumptions as the *chain ladder method*. Furthermore, in the examples

- the realizations of the grossing up predictors are identical with those of the *chain ladder predictors* and
- the realizations of the grossing up quotas are identical with those of the *chain ladder quotas*.

It turns out that these coincidences are not caused by rounding or hazard.

In fact, there are surprising relations between the grossing up quotas and the *chain ladder factors*

$$\varphi_k^{CL} := \frac{\sum_{j=0}^{n-k} S_{j,k}}{\sum_{j=0}^{n-k} S_{j,k-1}}$$

and hence between the grossing up quotas and the *chain ladder quotas*

$$\gamma_k^{CL} := \prod_{l=k+1}^{n} \frac{1}{\varphi_l^{CL}}$$

We have the following result:

Lemma. *The identity*

$$\gamma_{k-1}^{GU} \varphi_k^{CL} = \gamma_k^{GU}$$

holds for all $k \in \{1, \ldots, n\}$ and the identity

$$\gamma_k^{GU} = \gamma_k^{CL}$$

holds for all $k \in \{0, 1, \ldots, n\}$.

This lemma provides the basis for the following result:

Theorem. *The identity*

$$S_{i,k}^{GU} = S_{i,k}^{CL}$$

holds for all $i, k \in \{0, 1, \ldots, n\}$ such that $i + k \geq n + 1$.

Because of the theorem, the grossing up predictors are always identical with the *chain ladder predictors* of the future cumulative losses.

As the *chain ladder method* is more transparent than the recursively defined grossing up method, the grossing up method is irrelevant in practice.

Bornhuetter–Ferguson Principle

Because of the identity

$$S_{i,k}^{GU} = \gamma_k^{GU} \frac{S_{i,n-i}}{\gamma_{n-i}^{GU}}$$

the grossing up method is a special case of the *loss development method*. Moreover, letting

$$\alpha_i^{GU} := \frac{S_{i,n-i}}{\gamma_{n-i}^{GU}}$$

the grossing up predictors of the future cumulative losses can be represented in the form

$$S_{i,k}^{GU} = S_{i,n-i} + \left(\gamma_k^{GU} - \gamma_{n-i}^{GU}\right) \alpha_i^{GU}$$

Therefore, the grossing up method is subject to the *Bornhuetter–Ferguson principle*.

Notes

Keywords: Bornhuetter–Ferguson Method, Bornhuetter–Ferguson Principle, Development Patterns (Basics), Development Patterns (Estimation), Chain Ladder Method (Basics), Loss Development Method, Multiplicative Models, Run-Off Triangles.

References: Institute of Actuaries [1989], Lorenz & Schmidt [1999], Schmidt [2009].

Linear Models (Basics)

Kathrin Bach and Klaus D. Schmidt

Linear models belong to the basic models of mathematical statistics. They essentially consist of the assumption that the expectations of all random variables of the model depend on an unknown parameter vector via a known matrix.

In the elementary linear model all random variables are observable. In this model the problem is to estimate the parameter.

In the extended linear model some of the random variables are non-observable. In this model the problem is not only to estimate the parameter but also to predict the non-observable random variables. For both purposes, estimation of the parameter and prediction of the non-observable random variables, only the observable random variables are available.

In the extended linear model with a constraint, parameter estimation and prediction of the non-observable random variables has to be done under a linear constraint on the parameter.

Elementary Linear Model

Consider a random vector $\mathbf{X}$ with values in $\mathbb{R}^m$ and define

$$\Sigma := \mathrm{var}[\mathbf{X}]$$

K. Bach · K.D. Schmidt (✉)
Technische Universität Dresden, Dresden, Germany
e-mail: klaus.d.schmidt@tu-dresden.de

© Springer International Publishing Switzerland 2016
M. Radtke et al. (eds.), *Handbook on Loss Reserving*,
EAA Series, DOI 10.1007/978-3-319-30056-6_17

133

The elementary linear model for $\mathbf{X}$ consists of the following assumptions:

Elementary Linear Model:

(i) *There exists an unknown parameter $\beta \in \mathbb{R}^s$ and a known matrix $\mathbf{A} \subset \mathbb{R}^{m \times s}$ such that* rang$(\mathbf{A}) = s$ *and*

$$E[\mathbf{X}] = \mathbf{A}\beta$$

(ii) *The matrix Σ is positive definite.*

We assume in this section that the assumptions of the elementary linear model are fulfilled and that the matrix Σ is known.

In the elementary linear model the unknown parameter β has to be estimated.

In principle, every random vector $\widehat{\beta}$ with values in $\mathbb{R}^s$ which depends on hazard only via the random vector $\mathbf{X}$ is a possible *estimator* of β. For an estimator $\widehat{\beta}$ of β, the difference

$$\widehat{\beta} - \beta$$

is called the *estimation error* of $\widehat{\beta}$ and the expectation

$$E\left[(\widehat{\beta} - \beta)'(\widehat{\beta} - \beta)\right]$$

is called the *expected squared estimation error* of $\widehat{\beta}$. The expected squared estimation error provides the basis for the comparison of different estimators of β. Because of the identity

$$E\left[(\widehat{\beta}-\beta)'(\widehat{\beta}-\beta)\right] = \text{trace}\left(\text{var}[\widehat{\beta}-\beta]\right) + E[\widehat{\beta}-\beta]'E[\widehat{\beta}-\beta]$$
$$= \text{trace}\left(\text{var}[\widehat{\beta}]\right) + E[\widehat{\beta}-\beta]'E[\widehat{\beta}-\beta]$$

the expected squared estimation error of an estimator $\widehat{\beta}$ of β is determined by the expectation and the variance of the estimation error.

We consider now some possible properties of an estimator of β: An estimator $\widehat{\beta}$ with values in $\mathbb{R}^s$ is called

- an *admissible estimator* of β if there exists a matrix $\mathbf{Q}$ such that

$$\widehat{\beta} = \mathbf{QX}$$
$$\mathbf{I} = \mathbf{QA}_1$$

and it is called
- a *Gauss–Markov estimator* of β if it is an admissible estimator of β which minimizes the expected squared estimation error over all admissible estimators of β.

Every admissible estimator $\widehat{\beta}$ of β satisfies $E[\widehat{\beta}] = \beta$ (unbiasedness) and hence

$$E\left[(\widehat{\beta}-\beta)'(\widehat{\beta}-\beta)\right] = \text{trace}\left(\text{var}[\widehat{\beta}]\right)$$

Therefore, the expected squared estimation error of an admissible estimator of β is determined by its variance.

Gauss–Markov Theorem. *There exists a unique Gauss–Markov estimator β^{GM} of β and it satisfies*

$$\beta^{GM} = (\mathbf{A}'\Sigma^{-1}\mathbf{A})^{-1}\mathbf{A}'\Sigma^{-1}\mathbf{X}$$

and

$$\mathrm{var}[\beta^{GM}] = (\mathbf{A}'\Sigma^{-1}\mathbf{A})^{-1}$$

The Gauss–Markov estimator of β has another interesting property:

Lemma. *The Gauss–Markov estimator*

$$\beta^{GM} = (\mathbf{A}'\Sigma^{-1}\mathbf{A})^{-1}\mathbf{A}'\Sigma^{-1}\mathbf{X}$$

minimizes the weighted squared approximation error

$$(\mathbf{X} - \mathbf{A}\widehat{\beta})'\Sigma^{-1}(\mathbf{X} - \mathbf{A}\widehat{\beta})$$

over all random vectors $\widehat{\beta}$ with values in $\mathbb{R}^s$, and it is the only random vector having this property.

From the statistical point of view, minimization of the weighted squared approximation error by the Gauss–Markov estimator is without interest. However, in some cases the lemma simplifies the computation of the Gauss–Markov estimator; this is, in particular, the case when the matrix Σ is diagonal.

Extended Linear Model

Consider now a random vector $\mathbf{X}$ with values in $\mathbb{R}^m$ of which only the first m_1 coordinates are observable while the remaining $m_2 = m - m_1$ coordinates are non-observable. We write $\mathbf{X}$ as a block vector

$$\mathbf{X} = \begin{pmatrix} \mathbf{X}_1 \\ \mathbf{X}_2 \end{pmatrix}$$

and we assume that $\mathbf{X}_1$ is observable and that $\mathbf{X}_2$ is non-observable. Letting

$$\Sigma_{11} := \mathrm{var}[\mathbf{X}_1]$$
$$\Sigma_{12} := \mathrm{cov}[\mathbf{X}_1, \mathbf{X}_2]$$
$$\Sigma_{21} := \mathrm{cov}[\mathbf{X}_2, \mathbf{X}_1]$$
$$\Sigma_{22} := \mathrm{var}[\mathbf{X}_2]$$

we then obtain

$$\mathrm{var}[\mathbf{X}] = \begin{pmatrix} \Sigma_{11} & \Sigma_{12} \\ \Sigma_{21} & \Sigma_{22} \end{pmatrix}$$

We extend the linear model as follows:

Extended Linear Model:

(i) *There exist an unknown parameter $\beta \in \mathbb{R}^s$ and known matrices $\mathbf{A}_1 \in \mathbb{R}^{m_1 \times s}$ and $\mathbf{A}_2 \in \mathbb{R}^{m_2 \times s}$ such that* $\mathrm{rang}(\mathbf{A}_1) = s$ *and*

$$E\left[\begin{pmatrix} \mathbf{X}_1 \\ \mathbf{X}_2 \end{pmatrix}\right] = \begin{pmatrix} \mathbf{A}_1 \\ \mathbf{A}_2 \end{pmatrix}\beta$$

(ii) *The matrix Σ_{11} is positive definite.*

We assume now that the assumptions of the extended linear model are fulfilled and that the matrix Σ_{11} is known.

In the extended linear model the random vector $\mathbf{X}_2$ is non-observable. Therefore, only the random vector $\mathbf{X}_1$ is available for estimating the parameter β. Moreover, the problem arises to predict the random vector $\mathbf{X}_2$. It turns out that the estimation problem and the prediction problem can be handled in a completely analogous way.

Besides estimation of β and prediction of $\mathbf{X}_2$, the more general problems of estimating $\mathbf{C}\beta$ with $\mathbf{C} \in \mathbb{R}^{r \times s}$ and of predicting $\mathbf{D}\mathbf{X}_2$ with $\mathbf{D} \in \mathbb{R}^{r \times m_2}$ are of interest as well. Letting $\mathbf{C} = \mathbf{I}$ or $\mathbf{D} = \mathbf{I}$, respectively, we obtain the problems of estimating β and of predicting $\mathbf{X}_2$ as a special case. Of particular interest are also the cases $\mathbf{C} = \mathbf{c}'$ with $\mathbf{c} \in \mathbb{R}^s$ and $\mathbf{D} = \mathbf{d}'$ with $\mathbf{d} \in \mathbb{R}^{m_2}$, which are the problems of estimating or predicting a linear combination of the coordinates of β or $\mathbf{X}_2$, respectively.

Let us first study the estimation of $\mathbf{C}\beta$ with $\mathbf{C} \in \mathbb{R}^{r \times s}$.

In principle, every random vector $\widehat{\mathbf{Y}}$ with values in $\mathbb{R}^r$ which depends on hazard only via the observable random vector $\mathbf{X}_1$ is a possible *estimator* of $\mathbf{C}\beta$. For an estimator $\widehat{\mathbf{Y}}$ of $\mathbf{C}\beta$, the difference

$$\widehat{\mathbf{Y}} - \mathbf{C}\beta$$

is called the *estimation error* of $\widehat{\mathbf{Y}}$ and the expectation

$$E\left[(\widehat{\mathbf{Y}} - \mathbf{C}\beta)'(\widehat{\mathbf{Y}} - \mathbf{C}\beta)\right]$$

is called the *expected squared estimation error* of $\widehat{\mathbf{Y}}$. The expected squared estimation error provides the basis for the comparison of different estimators of $\mathbf{C}\beta$. Because of the identity

$$E\left[(\widehat{\mathbf{Y}} - \mathbf{C}\beta)'(\widehat{\mathbf{Y}} - \mathbf{C}\beta)\right] = \mathrm{trace}\left(\mathrm{var}[\widehat{\mathbf{Y}} - \mathbf{C}\beta]\right) + E[\widehat{\mathbf{Y}} - \mathbf{C}\beta]'E[\widehat{\mathbf{Y}} - \mathbf{C}\beta]$$
$$= \mathrm{trace}\left(\mathrm{var}[\widehat{\mathbf{Y}}]\right) + E[\widehat{\mathbf{Y}} - \mathbf{C}\beta]'E[\widehat{\mathbf{Y}} - \mathbf{C}\beta]$$

the expected squared estimation error of an estimator $\widehat{\mathbf{Y}}$ of $\mathbf{C}\beta$ is determined by the expectation and the variance of the estimation error.

We consider now some possible properties of an estimator of $\mathbf{C}\beta$: An estimator $\widehat{\mathbf{Y}}$ with values in $\mathbb{R}^r$ is called

- an *admissible estimator* of $\mathbf{C}\beta$ if there exists a matrix $\mathbf{Q}$ such that

$$\widehat{\mathbf{Y}} = \mathbf{Q}\mathbf{X}_1$$
$$\mathbf{C} = \mathbf{Q}\mathbf{A}_1$$

and it is called
- a *Gauss–Markov estimator* of $\mathbf{C}\beta$ if it is an admissible estimator of $\mathbf{C}\beta$ which minimizes the expected squared estimation error over all admissible estimators of $\mathbf{C}\beta$.

Every admissible estimator $\widehat{\mathbf{Y}}$ of $\mathbf{C}\beta$ satisfies $E[\widehat{\mathbf{Y}}] = \mathbf{C}\beta$ (unbiasedness) and hence

$$E\big[(\widehat{\mathbf{Y}} - \mathbf{C}\beta)'(\widehat{\mathbf{Y}} - \mathbf{C}\beta)\big] = \mathrm{trace}\big(\mathrm{var}[\widehat{\mathbf{Y}}]\big)$$

Therefore, the expected squared estimation error of an admissible estimator of $\mathbf{C}\beta$ is determined by its variance.

Gauss–Markov Theorem. *There exists a unique Gauss–Markov estimator* $(\mathbf{C}\beta)^{\mathrm{GM}}$ *of* $\mathbf{C}\beta$ *and it satisfies*

$$(\mathbf{C}\beta)^{\mathrm{GM}} = \mathbf{C}(\mathbf{A}'\Sigma_{11}^{-1}\mathbf{A}_1)^{-1}\mathbf{A}'\Sigma_{11}^{-1}\mathbf{X}_1$$

and

$$\mathrm{var}\big[(\mathbf{C}\beta)^{\mathrm{GM}}\big] = \mathbf{C}(\mathbf{A}'\Sigma_{11}^{-1}\mathbf{A}_1)^{-1}\mathbf{C}'$$

Because of the Gauss–Markov theorem, the Gauss–Markov estimator β^{GM} of β satisfies

$$\beta^{\mathrm{GM}} = (\mathbf{A}_1'\Sigma_{11}^{-1}\mathbf{A}_1)^{-1}\mathbf{A}_1'\Sigma_{11}^{-1}\mathbf{X}_1$$

and it now follows that the Gauss–Markov estimator of $\mathbf{C}\beta$ satisfies

$$(\mathbf{C}\beta)^{\mathrm{GM}} = \mathbf{C}\beta^{\mathrm{GM}}$$

This property is called *linearity* of Gauss–Markov estimation.

As in the elementary linear model, we have the following lemma:

Lemma. *The random vector*

$$\beta^{\mathrm{GM}} = (\mathbf{A}_1'\Sigma_{11}^{-1}\mathbf{A}_1)^{-1}\mathbf{A}_1'\Sigma_{11}^{-1}\mathbf{X}_1$$

minimizes the weighted squared approximation error

$$(\mathbf{X}_1 - \mathbf{A}_1\widehat{\beta})'\Sigma^{-1}(\mathbf{X}_1 - \mathbf{A}_1\widehat{\beta})$$

over all random vectors $\widehat{\beta}$ *with values in* $\mathbb{R}^s$, *and it is the only random vector having this property.*

Let us now study the prediction of $\mathbf{D}\mathbf{X}_2$ with $\mathbf{D} \in \mathbb{R}^{r \times m_2}$.

In principle, every random vector $\widehat{\mathbf{Y}}$ with values in $\mathbb{R}^r$ which depends on hazard only via the observable random vector $\mathbf{X}_1$ is a possible *predictor* of $\mathbf{DX}_2$. For a predictor $\widehat{\mathbf{Y}}$ of $\mathbf{DX}_2$, the difference

$$\widehat{\mathbf{Y}} - \mathbf{DX}_2$$

is called the *prediction error* of $\widehat{\mathbf{Y}}$ and the expectation

$$E\left[\left(\widehat{\mathbf{Y}} - \mathbf{DX}_2\right)'\left(\widehat{\mathbf{Y}} - \mathbf{DX}_2\right)\right]$$

is called the *expected squared prediction error* of $\widehat{\mathbf{Y}}$. The expected squared prediction error provides the basis for the comparison of different predictors of $\mathbf{DX}_2$. Because of the identity

$$E\left[\left(\widehat{\mathbf{Y}} - \mathbf{DX}_2\right)'\left(\widehat{\mathbf{Y}} - \mathbf{DX}_2\right)\right]$$
$$= \operatorname{trace}\left(\operatorname{var}\left[\widehat{\mathbf{Y}} - \mathbf{DX}_2\right]\right) + E\left[\widehat{\mathbf{Y}} - \mathbf{DX}_2\right]' E\left[\widehat{\mathbf{Y}} - \mathbf{DX}_2\right]$$

the expected squared prediction error of a predictor $\widehat{\mathbf{Y}}$ of $\mathbf{DX}_2$ is determined by the expectation and the variance of the prediction error.

We consider now some possible properties of a predictor of $\mathbf{DX}_2$: A predictor $\widehat{\mathbf{Y}}$ with values in $\mathbb{R}^r$ is called

- an *admissible predictor* of $\mathbf{DX}_2$ if there exists a matrix $\mathbf{Q}$ such that

$$\widehat{\mathbf{Y}} = \mathbf{Q}\mathbf{X}_1$$
$$\mathbf{DA}_2 = \mathbf{Q}\mathbf{A}_1$$

and it is called
- a *Gauss–Markov predictor* of $\mathbf{DX}_2$ if it is an admissible predictor of $\mathbf{DX}_2$ and minimizes the expected squared prediction error over all admissible predictors of $\mathbf{DX}_2$.

Every admissible predictor $\widehat{\mathbf{Y}}$ of $\mathbf{DX}_2$ satisfies $E[\widehat{\mathbf{Y}}] = E[\mathbf{DX}_2]$ (unbiasedness) and hence

$$E\left[\left(\widehat{\mathbf{Y}} - \mathbf{DX}_2\right)'\left(\widehat{\mathbf{Y}} - \mathbf{DX}_2\right)\right] = \operatorname{trace}\left(\operatorname{var}\left[\widehat{\mathbf{Y}} - \mathbf{DX}_2\right]\right)$$

Therefore, the expected squared prediction error of an admissible predictor of $\mathbf{DX}_2$ is determined by the variance of the prediction error.

Gauss–Markov Theorem. *There exists a unique Gauss–Markov predictor* $(\mathbf{DX}_2)^{\mathrm{GM}}$ *of* $\mathbf{DX}_2$ *and it satisfies*

$$(\mathbf{DX}_2)^{\mathrm{GM}} = \mathbf{D}\left(\mathbf{A}_2 \beta^{\mathrm{GM}} + \Sigma_{21} \Sigma_{11}^{-1}\left(\mathbf{X}_1 - \mathbf{A}_1 \beta^{\mathrm{GM}}\right)\right)$$

and

$$\text{var}\big[(\mathbf{DX}_2)^{\text{GM}} - \mathbf{DX}_2\big] = \mathbf{D}\Big(\big(\Sigma_{22} - \Sigma_{21}\Sigma_{11}^{-1}\Sigma_{12}\big)$$
$$+ \big(\mathbf{A}_2 - \Sigma_{21}\Sigma_{11}^{-1}\mathbf{A}_1\big)\big(\mathbf{A}_1'\Sigma_{11}^{-1}\mathbf{A}_1\big)^{-1}\big(\mathbf{A}_2 - \Sigma_{21}\Sigma_{11}^{-1}\mathbf{A}_1\big)'\Big)\mathbf{D}'$$

Because of the Gauss–Markov theorem, the Gauss–Markov predictor $\mathbf{X}_2^{\text{GM}}$ of $\mathbf{X}_2$ satisfies

$$\mathbf{X}_2^{\text{GM}} = \mathbf{A}_2\beta^{\text{GM}} + \Sigma_{21}\Sigma_{11}^{-1}(\mathbf{X}_1 - \mathbf{A}_1\beta^{\text{GM}})$$

and it then follows that

$$(\mathbf{DX}_2)^{\text{GM}} = \mathbf{DX}_2^{\text{GM}}$$

This property is called *linearity* of Gauss–Markov prediction. In particular, the co-ordinates of the Gauss–Markov predictor of $\mathbf{X}_2$ are precisely the Gauss–Markov predictors of the coordinates of $\mathbf{X}_2$.

We conclude this section with a special case:

Lemma. *Assume that* $\Sigma_{21} = \mathbf{O}$. *Then*

$$\mathbf{X}_2^{\text{GM}} = \mathbf{A}_2\beta^{\text{GM}}$$

and

$$\text{var}\big[\mathbf{X}_2^{\text{GM}} - \mathbf{X}_2\big] = \Sigma_{22} + \mathbf{A}_2(\mathbf{A}_1'\Sigma_{11}^{-1}\mathbf{A}_1)^{-1}\mathbf{A}_2'$$

This special case is of interest in the *additive model* of loss reserving.

Extended Linear Model with a Constraint

We maintain the notation of the preceding section and consider now the extended linear model with a linear constraint on the parameter.

Extended Linear Model with a Constraint:

(i) *There exist an unknown parameter* $\beta \in \mathbb{R}^s$ *and known matrices* $\mathbf{A}_1 \in \mathbb{R}^{m_1 \times s}$ *and* $\mathbf{A}_2 \in \mathbb{R}^{m_2 \times s}$ *such that* $\text{rang}(\mathbf{A}_1) = s$ *and*

$$E\left[\begin{pmatrix}\mathbf{X}_1 \\ \mathbf{X}_2\end{pmatrix}\right] = \begin{pmatrix}\mathbf{A}_1 \\ \mathbf{A}_2\end{pmatrix}\beta$$

(ii) *There exist a known vector* $\mathbf{h} \in \mathbb{R}^r$ *and a known matrix* $\mathbf{H} \in \mathbb{R}^{r \times s}$ *such that* $\text{rang}(\mathbf{H}) = r$ *and*

$$\mathbf{H}\beta = \mathbf{h}$$

(iii) *The matrix* Σ_{11} *is positive definite.*

An estimator $\widehat{\mathbf{Y}}$ of $\mathbf{C}\beta$ is called

- an $(\mathbf{H}, \mathbf{h})$-*admissible estimator* of $\mathbf{C}\beta$ if there exist matrices $\mathbf{Q}$ and $\mathbf{R}$ such that

$$\widehat{\mathbf{Y}} = \mathbf{Q}\mathbf{X}_1 + \mathbf{R}\mathbf{h}$$
$$\mathbf{C} = \mathbf{Q}\mathbf{A}_1 + \mathbf{R}\mathbf{H}$$

  and it is called
- an $(\mathbf{H}, \mathbf{h})$-*Gauss–Markov estimator* of $\mathbf{C}\beta$ if it is an $(\mathbf{H}, \mathbf{h})$-admissible estimator of $\mathbf{C}\beta$ which minimizes the expected squared estimation error over all $(\mathbf{H}, \mathbf{h})$-admissible estimators of $\mathbf{C}\beta$.

Correspondingly, a predictor $\widehat{\mathbf{Y}}$ of $\mathbf{D}\mathbf{X}_2$ is called

- an $(\mathbf{H}, \mathbf{h})$-*admissible predictor* of $\mathbf{D}\mathbf{X}_2$ if there exist matrices $\mathbf{Q}$ and $\mathbf{R}$ such that

$$\widehat{\mathbf{Y}} = \mathbf{Q}\mathbf{X}_1 + \mathbf{R}\mathbf{h}$$
$$\mathbf{D}\mathbf{A}_2 = \mathbf{Q}\mathbf{A}_1 + \mathbf{R}\mathbf{H}$$

  and it is called
- an $(\mathbf{H}, \mathbf{h})$-*Gauss–Markov predictor* of $\mathbf{D}\mathbf{X}_2$ if it is an $(\mathbf{H}, \mathbf{h})$-admissible predictor of $\mathbf{D}\mathbf{X}_2$ which minimizes the expected squared prediction error over all $(\mathbf{H}, \mathbf{h})$-admissible predictors of $\mathbf{D}\mathbf{X}_2$.

Every $(\mathbf{H}, \mathbf{h})$-admissible estimator of $\mathbf{C}\beta$ and every $(\mathbf{H}, \mathbf{h})$-admissible predictor of $\mathbf{D}\mathbf{X}_2$ is unbiased.

Gauss–Markov Theorem. *Define*

$$\mathbf{G} := (\mathbf{H}(\mathbf{A}_1' \Sigma_{11}^{-1} \mathbf{A}_1)^{-1}\mathbf{H}')^{-1}$$

There exists a unique $(\mathbf{H}, \mathbf{h})$-*Gauss–Markov estimator* $(\mathbf{C}\beta)_{\mathbf{H},\mathbf{h}}^{\mathrm{GM}}$ *of* $\mathbf{C}\beta$ *and it satisfies*

$$(\mathbf{C}\beta)_{\mathbf{H},\mathbf{h}}^{\mathrm{GM}} = \mathbf{C}\Big(\beta^{\mathrm{GM}} - (\mathbf{A}_1' \Sigma_{11}^{-1} \mathbf{A}_1)^{-1}\mathbf{H}'\mathbf{G}(\mathbf{H}\beta^{\mathrm{GM}} - \mathbf{h})\Big)$$

and there exists a unique $(\mathbf{H}, \mathbf{h})$-*Gauss–Markov predictor* $(\mathbf{D}\mathbf{X}_2)_{\mathbf{H},\mathbf{h}}^{\mathrm{GM}}$ *of* $\mathbf{D}\mathbf{X}_2$ *and it satisfies*

$$(\mathbf{D}\mathbf{X}_2)_{\mathbf{H},\mathbf{h}}^{\mathrm{GM}} = \mathbf{D}\Big(\mathbf{A}_1\beta_{\mathbf{H},\mathbf{h}}^{\mathrm{GM}} + \Sigma_{21}\Sigma_{11}^{-1}(\mathbf{X}_1 - \mathbf{A}_1\beta_{\mathbf{H},\mathbf{h}}^{\mathrm{GM}})\Big)$$

In the case $\mathbf{H}\beta^{\mathrm{GM}} = \mathbf{h}$ this yields $\beta_{\mathbf{H},\mathbf{h}}^{\mathrm{GM}} = \beta^{\mathrm{GM}}$ and hence $\mathbf{X}_{2;\mathbf{H},\mathbf{h}}^{\mathrm{GM}} = \mathbf{X}_2^{\mathrm{GM}}$.

Comparison with the Credibility Predictor

In this section we assume again that the assumptions of the extended linear model are fulfilled. Then the assumptions of the *extended credibility model* are fulfilled as well and the *credibility predictor* $\mathbf{X}_2^{\mathrm{CR}}$ of $\mathbf{X}_2$ satisfies

$$\mathbf{X}_2^{\mathrm{CR}} = \mathbf{A}_2\beta + \Sigma_{21}\Sigma_{11}^{-1}(\mathbf{X}_1 - \mathbf{A}_1\beta)$$

Comparison of the credibility predictor with the Gauss–Markov predictor

$$\mathbf{X}_2^{\mathrm{GM}} = \mathbf{A}_2 \beta^{\mathrm{GM}} + \Sigma_{21} \Sigma_{11}^{-1} (\mathbf{X}_1 - \mathbf{A}_1 \beta^{\mathrm{GM}})$$

shows that the Gauss–Markov predictor of $\mathbf{X}_2$ is obtained from the credibility predictor by replacing the unknown parameter β in the credibility predictor by its Gauss–Markov estimator β^{GM}.

This comparison provides an intuitive explanation of the fact that the expected squared prediction error of the Gauss–Markov predictor is at least as large as the expected squared prediction error of the credibility predictor. This is also evident from the fact that

- the credibility predictor of $\mathbf{X}_2$ minimizes the expected squared prediction error over all affine-linear predictors of $\mathbf{X}_2$ while
- the Gauss–Markov predictor of $\mathbf{X}_2$ minimizes the expected squared prediction error over all admissible (and hence unbiased and linear) predictors of $\mathbf{X}_2$.

These two predictors are identical only under a very particular condition on the matrices $\mathbf{A}_1$ and $\mathbf{A}_2$:

Theorem. *The following properties are equivalent:*

(a) *The matrices $\mathbf{A}_1$ and $\mathbf{A}_2$ satisfy $\mathbf{A}_2 - \Sigma_{21} \Sigma_{11}^{-1} \mathbf{A}_1 = \mathbf{O}$.*

(b) *The Gauss–Markov predictor of $\mathbf{X}_2$ is identical with the credibility predictor of $\mathbf{X}_2$.*

(c) *The expected squared prediction error of the Gauss–Markov predictor of $\mathbf{X}_2$ is identical with the expected squared prediction error of the credibility predictor of $\mathbf{X}_2$.*

These observations clarify the relation between the Gauss–Markov predictor and the credibility predictor.

Remark

For a given sub-σ-algebra $\mathcal{G}$ of the probability space, replacing the first and second moments of the random vectors in the extended linear model by the corresponding $\mathcal{G}$-*conditional moments* and also replacing the matrices $\mathbf{A}_1$ and $\mathbf{A}_2$ by $\mathcal{G}$-measurable random matrices leads to a $\mathcal{G}$-*conditional linear model* and accordingly to $\mathcal{G}$-*conditional Gauss–Markov estimators* and *predictors*. Correspondingly, a $\mathcal{G}$-conditional version of the extended linear model with a constraint is obtained by also replacing the matrix $\mathbf{H}$ by a $\mathcal{G}$-measurable random matrix and the vector $\mathbf{h}$ by a $\mathcal{G}$-measurable random vector.

Notes

Keywords: Credibility Models (Basics), Linear Models (Loss Reserving).

References: Hamer [1999], Kloberdanz & Schmidt [2008], Ludwig, Schmeißer & Thänert [2009], Schmidt [1998, 1999c, 2012].

Linear Models (Loss Reserving)

Klaus D. Schmidt

Linear models and Gauss–Markov predictors are also useful in loss reserving. Some particularities arise from the structure of the run-off square and from the task to determine Gauss–Markov predictors of different reserves. We explain these particularities and then discuss a special linear model for loss reserving.

Consider the run-off square of incremental losses:

Accident year	Development year					
	0	1	... k	... $n-i$	... $n-1$	n
0	$Z_{0,0}$	$Z_{0,1}$	... $Z_{0,k}$	... $Z_{0,n-i}$	... $Z_{0,n-1}$	$Z_{0,n}$
1	$Z_{1,0}$	$Z_{1,1}$	... $Z_{1,k}$	... $Z_{1,n-i}$	... $Z_{1,n-1}$	$Z_{1,n}$
$\vdots$	$\vdots$	$\vdots$	$\vdots$	$\vdots$	$\vdots$	$\vdots$
i	$Z_{i,0}$	$Z_{i,1}$	... $Z_{i,k}$	... $Z_{i,n-i}$	... $Z_{i,n-1}$	$Z_{i,n}$
$\vdots$	$\vdots$	$\vdots$	$\vdots$	$\vdots$	$\vdots$	$\vdots$
$n-k$	$Z_{n-k,0}$	$Z_{n-k,1}$	... $Z_{n-k,k}$	... $Z_{n-k,n-i}$	... $Z_{n-k,n-1}$	$Z_{n-k,n}$
$\vdots$	$\vdots$	$\vdots$	$\vdots$	$\vdots$	$\vdots$	$\vdots$
$n-1$	$Z_{n-1,0}$	$Z_{n-1,1}$	... $Z_{n-1,k}$	... $Z_{n-1,n-i}$	... $Z_{n-1,n-1}$	$Z_{n-1,n}$
n	$Z_{n,0}$	$Z_{n,1}$	... $Z_{n,k}$	... $Z_{n,n-i}$	... $Z_{n,n-1}$	$Z_{n,n}$

We assume that the incremental losses $Z_{i,k}$ are observable for $i + k \leq n$ and that they are non-observable for $i + k \geq n + 1$.

Let $m := (n + 1)^2$ as well as $m_1 := (n + 1)(n + 2)/2$ and $m_2 := n(n + 1)/2$. We represent the incremental losses $Z_{i,k}$ as a random vector

K.D. Schmidt (✉)
Technische Universität Dresden, Dresden, Germany
e-mail: klaus.d.schmidt@tu-dresden.de

© Springer International Publishing Switzerland 2016
M. Radtke et al. (eds.), *Handbook on Loss Reserving*,
EAA Series, DOI 10.1007/978-3-319-30056-6_18

$$\mathbf{X} = \begin{pmatrix} \mathbf{X}_1 \\ \mathbf{X}_2 \end{pmatrix}$$

with values in $\mathbb{R}^m$ such that $\mathbf{X}_1$ consists of the m_1 observable incremental losses and $\mathbf{X}_2$ consists of the m_2 non-observable incremental losses.

In order to obtain a linear model for the incremental losses we assume that there exist an unknown parameter $\beta \in \mathbb{R}^s$ with $s \leq m$ and known vectors $\mathbf{g}_{i,k} \in \mathbb{R}^s$ such that the identity

$$E[Z_{i,k}] = \mathbf{g}'_{i,k}\beta$$

holds for all $i, k \in \{0, 1, \ldots, n\}$. Arranging the row vectors $\mathbf{g}'_{i,k}$ in the same way as the incremental losses yields a matrix

$$\mathbf{A} = \begin{pmatrix} \mathbf{A}_1 \\ \mathbf{A}_2 \end{pmatrix}$$

such that

$$E\left[\begin{pmatrix} \mathbf{X}_1 \\ \mathbf{X}_2 \end{pmatrix}\right] = \begin{pmatrix} \mathbf{A}_1 \\ \mathbf{A}_2 \end{pmatrix}\beta$$

If $\mathrm{rang}(\mathbf{A}_1) = s$ and $\mathrm{var}[\mathbf{X}_1]$ is positive definite, then the assumptions of the extended linear model are fulfilled.

We are interested in

- the *Gauss–Markov predictors*

$$\left(\sum_{l=n-i+1}^{n} Z_{i,l}\right)^{\mathrm{GM}}$$

of the accident year reserves $\sum_{l=n-i+1}^{n} Z_{i,l}$ with $i \in \{1, \ldots, n\}$,
- the *Gauss–Markov predictors*

$$\left(\sum_{l=c-n}^{n} Z_{c-l,l}\right)^{\mathrm{GM}}$$

of the calendar year reserves $\sum_{l=c-n}^{n} Z_{c-l,l}$ with $c \in \{n+1, \ldots, 2n\}$ and
- the *Gauss–Markov predictors*

$$\left(\sum_{l=1}^{n}\sum_{j=n-l+1}^{n} Z_{j,l}\right)^{\mathrm{GM}}$$

of the aggregate loss reserve $\sum_{l=1}^{n}\sum_{j=n-l+1}^{n} Z_{j,l}$.

These Gauss–Markov predictors are also called *Gauss–Markov reserves*. In all three cases we have to determine the Gauss–Markov predictor of a random variable of the form $\mathbf{d}'\mathbf{X}_2$ with some $\mathbf{d} \in \mathbb{R}^{m_2}$, and we know that

$$(\mathbf{d}'\mathbf{X}_2)^{\mathrm{GM}} = \mathbf{d}'\mathbf{X}_2^{\mathrm{GM}}$$

Furthermore, since the coordinates of the Gauss–Markov predictor of $\mathbf{X}_2$ are identical with the Gauss–Markov predictors of the coordinates of $\mathbf{X}_2$, we obtain

$$\left(\sum_{l=n-i+1}^{n} Z_{i,l} \right)^{\mathrm{GM}} = \sum_{l=n-i+1}^{n} Z_{i,l}^{\mathrm{GM}}$$

$$\left(\sum_{l=c-n}^{n} Z_{c-l,l} \right)^{\mathrm{GM}} = \sum_{l=c-n}^{n} Z_{c-l,l}^{\mathrm{GM}}$$

$$\left(\sum_{l=1}^{n} \sum_{j=n-l+1}^{n} Z_{j,l} \right)^{\mathrm{GM}} = \sum_{l=1}^{n} \sum_{j=n-l+1}^{n} Z_{j,l}^{\mathrm{GM}}$$

To determine the Gauss–Markov reserves it is thus sufficient to determine the Gauss–Markov predictors of the future incremental losses.

Additive Model

The additive model consists of the following assumptions:

Additive Model: *There exist known volume measures $v_0, v_1, \ldots, v_n$ of the accident years as well as unknown parameters $\zeta_0, \zeta_1, \ldots, \zeta_n$ and parameters $\sigma_0^2, \sigma_1^2, \ldots, \sigma_n^2$ such that the identities*

$$E\left[\frac{Z_{i,k}}{v_i} \right] = \zeta_k$$

$$\mathrm{cov}\left[\frac{Z_{i,k}}{v_i}, \frac{Z_{j,l}}{v_j} \right] = \frac{\sigma_k^2}{v_i} \delta_{i,j} \delta_{k,l}$$

hold for all $i, j, k, l \in \{0, 1, \ldots, n\}$.

In the additive model the parameters $\zeta_0, \zeta_1, \ldots, \zeta_n$ form a *development pattern for incremental loss ratios*. Moreover, the additive model is a *multiplicative model*.

In what follows we assume that the conditions of the additive model are fulfilled. Initially, we also assume that the parameters $\sigma_0^2, \sigma_1^2, \ldots, \sigma_n^2$ are known, but this assumption will be dropped later.

The conditions of the additive model yield

$$E[Z_{i,k}] = v_i \zeta_k$$

$$\mathrm{cov}[Z_{i,k}, Z_{j,l}] = v_i \sigma_k^2 \delta_{i,j} \delta_{k,l}$$

and letting

$$\beta := \begin{pmatrix} \zeta_0 \\ \zeta_1 \\ \vdots \\ \zeta_n \end{pmatrix}$$

and

$$\mathbf{g}_{i,k} := v_i \, \mathbf{e}_{k+1}$$

we obtain $s = n + 1$ and

$$E[Z_{i,k}] = \mathbf{g}'_{i,k}\boldsymbol{\beta}$$

Therefore, there exist matrices $\mathbf{A}_1 \in \mathbb{R}^{m_1 \times s}$ and $\mathbf{A}_2 \in \mathbb{R}^{m_2 \times s}$ such that

$$E\left[\begin{pmatrix} \mathbf{X}_1 \\ \mathbf{X}_2 \end{pmatrix}\right] = \begin{pmatrix} \mathbf{A}_1 \\ \mathbf{A}_2 \end{pmatrix}\beta$$

and there exist diagonal matrices $\Sigma_{11} \in \mathbb{R}^{m_1 \times m_1}$ and $\Sigma_{22} \in \mathbb{R}^{m_2 \times m_2}$ such that

$$\mathrm{var}\left[\begin{pmatrix} \mathbf{X}_1 \\ \mathbf{X}_2 \end{pmatrix}\right] = \begin{pmatrix} \Sigma_{11} & \mathbf{O} \\ \mathbf{O} & \Sigma_{22} \end{pmatrix}\beta$$

Moreover, we have $\mathrm{rang}(\mathbf{A}_1) = s$ and $\Sigma_{11} = \mathrm{var}[\mathbf{X}_1]$ is invertible. Therefore, the additive model is an *extended linear model* with $\Sigma_{21} = \mathrm{cov}[\mathbf{X}_2, \mathbf{X}_1] = \mathbf{O}$.

Example. Let $n = 2$. Then we have $m_1 = 6$ and $m_2 = 3$ as well as $s = 3$, and we obtain

$$E\left[\begin{pmatrix} Z_{0,0} \\ Z_{0,1} \\ Z_{0,2} \\ Z_{1,0} \\ Z_{1,1} \\ Z_{2,0} \\ \hline Z_{1,2} \\ Z_{2,1} \\ Z_{2,2} \end{pmatrix}\right] = \begin{pmatrix} v_0 & 0 & 0 \\ 0 & v_0 & 0 \\ 0 & 0 & v_0 \\ v_1 & 0 & 0 \\ 0 & v_1 & 0 \\ v_2 & 0 & 0 \\ \hline 0 & 0 & v_1 \\ 0 & v_2 & 0 \\ 0 & 0 & v_2 \end{pmatrix} \begin{pmatrix} \zeta_0 \\ \zeta_1 \\ \zeta_2 \end{pmatrix}$$

and

$$
\mathrm{var}\left[\begin{pmatrix} Z_{0,0} \\ Z_{0,1} \\ Z_{0,2} \\ Z_{1,0} \\ Z_{1,1} \\ Z_{2,0} \\ \hline Z_{1,2} \\ Z_{2,1} \\ Z_{2,2} \end{pmatrix}\right] = \left(\begin{array}{cccccc|ccc}
v_0\,\sigma_0^2 & 0 & 0 & 0 & 0 & 0 & 0 & 0 & 0 \\
0 & v_0\,\sigma_1^2 & 0 & 0 & 0 & 0 & 0 & 0 & 0 \\
0 & 0 & v_0\,\sigma_2^2 & 0 & 0 & 0 & 0 & 0 & 0 \\
0 & 0 & 0 & v_1\,\sigma_0^2 & 0 & 0 & 0 & 0 & 0 \\
0 & 0 & 0 & 0 & v_1\,\sigma_1^2 & 0 & 0 & 0 & 0 \\
0 & 0 & 0 & 0 & 0 & v_2\,\sigma_0^2 & 0 & 0 & 0 \\
\hline
0 & 0 & 0 & 0 & 0 & 0 & v_1\,\sigma_2^2 & 0 & 0 \\
0 & 0 & 0 & 0 & 0 & 0 & 0 & v_2\,\sigma_1^2 & 0 \\
0 & 0 & 0 & 0 & 0 & 0 & 0 & 0 & v_2\,\sigma_2^2
\end{array}\right)
$$

Therefore the conditions of the extended linear model are fulfilled.

Since Σ_{11} is a diagonal matrix and the Gauss–Markov estimator β^{GM} is the unique minimizer of the *weighted squared approximation error*

$$
\left(\mathbf{X}_1 - \mathbf{A}_1\widehat{\beta}\right)'\Sigma_{11}^{-1}\left(\mathbf{X}_1 - \mathbf{A}_1\widehat{\beta}\right)
$$

over all random vectors $\widehat{\beta}$ with values in $\mathbb{R}^s$, the Gauss–Markov estimator β^{GM} can be determined in a particularly simple way: With

$$
\widehat{\beta} = \begin{pmatrix} \widehat{\zeta_0} \\ \widehat{\zeta_1} \\ \vdots \\ \widehat{\zeta_n} \end{pmatrix}
$$

and using the definitions of $\mathbf{X}_1$ and $\mathbf{A}_1$ we obtain

$$
\left(\mathbf{X}_1 - \mathbf{A}_1\widehat{\beta}\right)'\Sigma_{11}^{-1}\left(\mathbf{X}_1 - \mathbf{A}_1\widehat{\beta}\right) = \sum_{l=0}^{n}\sum_{j=0}^{n-l}\frac{1}{v_j\,\sigma_l^2}\left(Z_{j,l} - v_j\,\widehat{\zeta_l}\right)^2
$$

$$
= \sum_{l=0}^{n}\frac{1}{\sigma_l^2}\sum_{j=0}^{n-l}\frac{1}{v_j}\left(Z_{j,l} - v_j\,\widehat{\zeta_l}\right)^2
$$

Therefore, minimization of the weighted squared approximation error can be achieved by minimization of the inner sums of the last expression and it follows that the Gauss–Markov estimator β^{GM} satisfies

$$
\beta^{\mathrm{GM}} = \begin{pmatrix} \zeta_0^{\mathrm{GM}} \\ \zeta_1^{\mathrm{GM}} \\ \vdots \\ \zeta_n^{\mathrm{GM}} \end{pmatrix}
$$

with

$$\zeta_k^{\mathrm{GM}} = \frac{\sum_{j=0}^{n-k} Z_{j,k}}{\sum_{j=0}^{n-k} v_j}$$

Furthermore, the identity

$$\mathrm{cov}[\zeta_k^{\mathrm{GM}}, \zeta_l^{\mathrm{GM}}] = \frac{\sigma_k^2}{\sum_{j=0}^{n-k} v_j} \delta_{k,l}$$

holds for all $k, l \in \{0, 1, \ldots, n\}$, which means that the coordinates of the Gauss–Markov estimator β^{GM} are uncorrelated.

Since $\Sigma_{21} = \mathbf{O}$ the Gauss–Markov predictor $\mathbf{X}_2^{\mathrm{GM}}$ satisfies

$$\mathbf{X}_2^{\mathrm{GM}} = \mathbf{A}_2 \beta^{\mathrm{GM}}$$

and since the Gauss–Markov predictors $Z_{i,k}^{\mathrm{GM}}$ of the future incremental losses $Z_{i,k}$ with $i + k \geq n + 1$ are exactly the coordinates of the Gauss–Markov predictor $\mathbf{X}_2^{\mathrm{GM}}$ we obtain

$$Z_{i,k}^{\mathrm{GM}} = v_i \, \zeta_k^{\mathrm{GM}}$$

Furthermore, the identity

$$\mathrm{cov}[Z_{i,k}^{\mathrm{GM}} - Z_{i,k}, Z_{j,l}^{\mathrm{GM}} - Z_{j,l}] = v_i v_j \left(\frac{1}{\sum_{h=0}^{n-k} v_h} + \frac{1}{v_i} \delta_{i,j} \right) \sigma_k^2 \, \delta_{k,l}$$

holds for all $i, j, k, l \in \{0, 1, \ldots, n\}$ such that $i + k \geq n + 1$ and $j + l \geq n + 1$. In particular, the identities

$$E[(Z_{i,k}^{\mathrm{GM}} - Z_{i,k})^2] = \mathrm{var}[Z_{i,k}^{\mathrm{GM}} - Z_{i,k}] = v_i^2 \left(\frac{1}{\sum_{h=0}^{n-k} v_h} + \frac{1}{v_i} \right) \sigma_k^2$$

yield the expected squared prediction errors of the future incremental losses.

Now all Gauss–Markov reserves are obtained by summation from the Gauss–Markov predictors of the future incremental losses, and the expected squared prediction errors of the Gauss–Markov reserves are obtained from the covariances of the prediction errors of the Gauss–Markov predictors of the future incremental losses.

Let us finally drop the unrealistic assumption that the variance parameters $\sigma_0^1, \sigma_1^2, \ldots, \sigma_n^2$ are known. Then these parameters have to be estimated. For $k \in \{0, 1, \ldots, n-1\}$,

$$\widehat{\sigma}_k^2 := \frac{1}{n-k} \sum_{j=0}^{n-k} v_j \left(\frac{Z_{j,k}}{v_j} - \zeta_k^{\mathrm{GM}} \right)^2$$

is an unbiased estimator of σ_k^2, and from these estimators an estimator of σ_n^2 can be determined by extrapolation.

The additive model provides the basis for the *additive method*.

Remarks

Along the lines of the discussion of the additive model it can be shown that the *Panning model*, which provides the basis for the *Panning method*, is a conditional linear model.

Furthermore, the *chain ladder method* can at least partially be justified by a sequential conditional linear model, consisting of a conditional linear model for each of the development years $k \in \{1, \ldots, n\}$ and using cumulative instead of incremental losses.

Notes

Keywords: Additive Method, Chain Ladder Method (Models), Development Patterns (Basics), Linear Models (Basics), Multiplicative Models, Multivariate Methods, Munich Chain Ladder Method, Paid & Incurred Problem, Panning Method, Volume Measures.

References: Ludwig, Schmeißer & Thänert [2009, 2011], Mack [2002], Schmidt [2012].

Lognormal Loglinear Model (Basics)

Klaus D. Schmidt

In the lognormal loglinear model the distribution of some random vector is specified by assuming that its logarithm, i.e. the random vector consisting of the logarithms of the coordinates of the given random vector, has a normal distribution and fulfills the conditions of a linear model.

We first study the elementary lognormal loglinear model and then pass to the extended lognormal loglinear model. We present an unbiased predictor of a linear combination of the coordinates of the given random vector, and hence an estimator of the expectation of this linear combination, as well as an unbiased estimator of the variance of that predictor or estimator.

Elementary Lognormal Loglinear Model

We consider a random vector $\mathbf{Z}$ with values in $(0, \infty)^m$ and we denote by $\ln(\mathbf{Z})$ the random vector whose coordinates are the logarithms of the coordinates of $\mathbf{Z}$.

The elementary lognormal loglinear model for $\mathbf{Z}$ consists of the following assumptions:

> **Elementary Lognormal Loglinear Model**: *There exists an unknown parameter $\beta \in \mathbb{R}^s$ as well as a known matrix $\mathbf{A} \in \mathbb{R}^{m \times s}$ with $\mathrm{rang}(\mathbf{A}) = s$ and a positive definite symmetric matrix $\Sigma \in \mathbb{R}^{m \times m}$ such that $P_{\mathbf{Z}} = \mathbf{LN}(\mathbf{A}\beta, \Sigma)$.*

We assume in this section that the assumptions of the elementary lognormal loglinear model are fulfilled and that the matrix Σ is known.

K.D. Schmidt (✉)
Technische Universität Dresden, Dresden, Germany
e-mail: klaus.d.schmidt@tu-dresden.de

© Springer International Publishing Switzerland 2016
M. Radtke et al. (eds.), *Handbook on Loss Reserving*,
EAA Series, DOI 10.1007/978-3-319-30056-6_19

Since the condition $P_{\mathbf{Z}} = \mathbf{LN}(\mathbf{A}\beta, \Sigma)$ is equivalent to $P_{\ln(\mathbf{Z})} = \mathbf{N}(\mathbf{A}\beta, \Sigma)$, the elementary lognormal loglinear model for $\mathbf{Z}$ yields an elementary linear model for the random vector $\ln(\mathbf{Z})$. The *Gauss–Markov estimator* β^{GM} of β is given by

$$\beta^{\mathrm{GM}} = (\mathbf{A}'\Sigma^{-1}\mathbf{A})^{-1}\mathbf{A}'\Sigma^{-1}\ln(\mathbf{Z})$$

and the assumption $P_{\ln(\mathbf{Z})} = \mathbf{N}(\mathbf{A}\beta, \Sigma)$ yields

$$P_{\beta^{\mathrm{GM}}} = \mathbf{N}(\beta, (\mathbf{A}'\Sigma^{-1}\mathbf{A})^{-1})$$

Furthermore, since the Gauss–Markov estimator of β minimizes the *weighted squared approximation error*

$$\big(\ln(\mathbf{Z}) - \mathbf{A}\widehat{\beta}\big)' \Sigma^{-1} \big(\ln(\mathbf{Z}) - \mathbf{A}\widehat{\beta}\big)$$

over all random vectors $\widehat{\beta}$ with values in $\mathbb{R}^s$, it maximizes the likelihood function

$$L(\widehat{\beta}) := \frac{1}{\sqrt{(2\pi)^m \det(\Sigma)} \prod_{i=1}^m Z_i} \exp\left(-\frac{1}{2}\big(\ln(\mathbf{Z}) - \mathbf{A}\widehat{\beta}\big)' \Sigma^{-1} \big(\ln(\mathbf{Z}) - \mathbf{A}\widehat{\beta}\big)\right)$$

Therefore, the Gauss–Markov estimator of β is also a *maximum likelihood estimator* of β.

Since the parameter β has no immediate interpretation with respect to the random vector $\mathbf{Z}$, estimation of β as such is of minor interest. However, the Gauss–Markov estimator of β serves as an auxiliary tool to construct unbiased predictors of the coordinates of $\mathbf{Z}$, and hence unbiased estimators of their expectations.

For $\mathbf{c} \in \mathbb{R}^m$ and $\mathbf{C} \in \mathbb{R}^{m \times m}$ we consider the transformation

$$h(\beta; \mathbf{c}, \mathbf{C}) := \exp\left(\mathbf{c}'\mathbf{A}\beta + \frac{1}{2}\operatorname{trace}(\Sigma\mathbf{C})\right)$$

of the parameter β. To estimate $h(\beta; \mathbf{c}, \mathbf{C})$ one might be tempted to use the random variable

$$h(\beta^{\mathrm{GM}}; \mathbf{c}, \mathbf{C}) = \exp\left(\mathbf{c}'\mathbf{A}\beta^{\mathrm{GM}} + \frac{1}{2}\operatorname{trace}(\Sigma\mathbf{C})\right)$$

but in general this estimator fails to be unbiased since

$$E[h(\beta^{\mathrm{GM}}; \mathbf{c}, \mathbf{C})] = h(\beta; \mathbf{c}, \mathbf{C})\exp\left(\frac{1}{2}\mathbf{c}'\mathbf{A}(\mathbf{A}'\Sigma^{-1}\mathbf{A})^{-1}\mathbf{A}'\mathbf{c}\right)$$

Because of this identity, the random variable

$$H(\beta^{\mathrm{GM}}; \mathbf{c}, \mathbf{C}) := \exp\left(\mathbf{c}'\mathbf{A}\beta^{\mathrm{GM}} + \frac{1}{2}\operatorname{trace}(\Sigma\mathbf{C}) - \frac{1}{2}\mathbf{c}'\mathbf{A}(\mathbf{A}'\Sigma^{-1}\mathbf{A})^{-1}\mathbf{A}'\mathbf{c}\right)$$

is an unbiased estimator of $h(\beta; \mathbf{c}, \mathbf{C})$. Moreover, it can be shown that the random variable

$$H(\beta^{\mathrm{GM}}; \mathbf{c}_1, \mathbf{C}_1)\,H(\beta^{\mathrm{GM}}; \mathbf{c}_2, \mathbf{C}_2) - H(\beta^{\mathrm{GM}}; \mathbf{c}_1 + \mathbf{c}_2, \mathbf{C}_1 + \mathbf{C}_2)$$

is an unbiased estimator of $\operatorname{cov}[H(\beta^{\mathrm{GM}}; \mathbf{c}_1, \mathbf{C}_1), H(\beta^{\mathrm{GM}}; \mathbf{c}_2, \mathbf{C}_2)]$.

We consider now a linear combination of the coordinates of $\mathbf{Z}$. First, for every $i \in \{1, \dots, m\}$ and with $\mathbf{E}_i := \mathbf{e}_i\mathbf{e}_i'$, we have

$$E[H(\beta^{\mathrm{GM}}; \mathbf{e}_i, \mathbf{E}_i)] = h(\beta; \mathbf{e}_i, \mathbf{E}_i) = \exp\left(\mathbf{e}_i'\mathbf{A}\beta + \frac{1}{2}\mathbf{e}_i'\Sigma\mathbf{e}_i\right) = E[Z_i]$$

For every $\mathbf{d} \in \mathbb{R}^m$, this yields

$$E\left[\sum_{i=1}^m d_i\, H(\beta^{\mathrm{GM}}; \mathbf{e}_i, \mathbf{E}_i)\right] = E\left[\sum_{i=1}^m d_i\, Z_i\right] = E[\mathbf{d}'\mathbf{Z}]$$

which means that

$$\sum_{i=1}^m d_i\, H(\beta^{\mathrm{GM}}; \mathbf{e}_i, \mathbf{E}_i)$$

is an unbiased predictor of $\mathbf{d}'\mathbf{Z}$ and hence an unbiased estimator of $E[\mathbf{d}'\mathbf{Z}]$. Moreover, it can be shown that

$$\sum_{i=1}^m \sum_{j=1}^m d_i d_j \Big(H(\beta^{\mathrm{GM}}; \mathbf{e}_i, \mathbf{E}_i)\,H(\beta^{\mathrm{GM}}; \mathbf{e}_j, \mathbf{E}_j) - H(\beta^{\mathrm{GM}}; \mathbf{e}_i + \mathbf{e}_j, \mathbf{E}_i + \mathbf{E}_j)\Big)$$

is an unbiased estimator of the variance of $\sum_{i=1}^m d_i\, H(\beta^{\mathrm{GM}}; \mathbf{e}_i, \mathbf{E}_i)$. It is thus possible to specify confidence intervals for the expectation of $\mathbf{d}'\mathbf{Z}$.

Extended Lognormal Loglinear Model

We consider now a random vector $\mathbf{Z}$ with values in $(0, \infty)^m$ of which only the first m_1 coordinates are observable while the remaining $m_2 = m - m_1$ coordinates are non-observable. We write $\mathbf{Z}$ as a block vector

$$\mathbf{Z} = \begin{pmatrix} \mathbf{Z}_1 \\ \mathbf{Z}_2 \end{pmatrix}$$

and we assume that $\mathbf{Z}_1$ is observable and that $\mathbf{Z}_2$ is non-observable. Letting

$$\Sigma_{ik} := \operatorname{cov}[\ln(\mathbf{Z}_i), \ln(\mathbf{Z}_k)]$$

for $i, k \in \{1, 2\}$ and

$$\Sigma := \begin{pmatrix} \Sigma_{11} & \Sigma_{12} \\ \Sigma_{21} & \Sigma_{22} \end{pmatrix}$$

we then obtain

$$\operatorname{var}[\ln(\mathbf{Z})] = \Sigma$$

We extend the lognormal loglinear model as follows:

Extended Lognormal Loglinear Model: *There exist an unknown parameter $\beta \in \mathbb{R}^s$ as well as a known matrix*

$$\mathbf{A} = \begin{pmatrix} \mathbf{A}_1 \\ \mathbf{A}_2 \end{pmatrix}$$

with $\mathbf{A}_1 \in \mathbb{R}^{m_1 \times s}$ and $\mathbf{A}_2 \in \mathbb{R}^{m_2 \times s}$ as well as $\operatorname{rang}(\mathbf{A}_1) = s$ such that $P_{\mathbf{Z}} = \mathbf{LN}(\mathbf{A}\beta, \Sigma)$.

We assume now that the assumptions of the extended lognormal loglinear model are fulfilled and that the matrix Σ_{11} is known.

The results of the elementary lognormal loglinear model can be transferred to the extended lognormal loglinear model with a minor modification: The Gauss–Markov estimator β^{GM} of β is given by

$$\beta^{\mathrm{GM}} = (\mathbf{A}_1' \Sigma_{11}^{-1} \mathbf{A}_1)^{-1} \mathbf{A}_1' \Sigma_{11}^{-1} \ln(\mathbf{Z}_1)$$

and the assumption $P_{\ln(\mathbf{Z}_1)} = \mathbf{N}(\mathbf{A}_1\beta, \Sigma_{11})$ yields

$$P_{\beta^{\mathrm{GM}}} = \mathbf{N}(\beta, (\mathbf{A}_1' \Sigma_{11}^{-1} \mathbf{A}_1)^{-1})$$

For $\mathbf{c} \in \mathbb{R}^m$ and $\mathbf{C} \in \mathbb{R}^{m \times m}$, it then follows that

$$H(\beta^{\mathrm{GM}}; \mathbf{c}, \mathbf{C}) := \exp\left(\mathbf{c}'\mathbf{A}\beta^{\mathrm{GM}} + \frac{1}{2}\operatorname{trace}(\Sigma\mathbf{C}) - \frac{1}{2}\mathbf{c}'\mathbf{A}(\mathbf{A}_1' \Sigma_{11}^{-1} \mathbf{A}_1)^{-1}\mathbf{A}'\mathbf{c}\right)$$

is an unbiased estimator of

$$h(\beta; \mathbf{c}, \mathbf{C}) := \exp\left(\mathbf{c}'\mathbf{A}\beta + \frac{1}{2}\operatorname{trace}(\Sigma\mathbf{C})\right)$$

Using these estimators it is then possible to construct, as in the elementary lognormal loglinear model, an unbiased predictor of a linear combination $\mathbf{d}'\mathbf{Z}$ of the coordinates of $\mathbf{Z}$ or an unbiased estimator of its expectation $E[\mathbf{d'Z}]$, as well as an unbiased estimator of the variance of these predictors or estimators.

In view of reserves, linear combinations involving only coordinates of the non-observable random vector $\mathbf{Z}_2$ are of particular interest in the extended lognormal loglinear model.

Notes

Keywords: Linear Models (Basics), Linear Models (Loss Reserving), Lognormal Loglinear Model (Loss Reserving).

Reference: Ludwig, Schmeißer & Thänert [2009].

Lognormal Loglinear Model
(Loss Reserving)

Stefan Kaulfuß, Klaus D. Schmidt and Katrin Thänert

The lognormal loglinear model can, in particular, be applied in loss reserving. In this article we first discuss the particularities arising in this application and we then study a special lognormal loglinear model in loss reserving.

Consider the run-off square of incremental losses:

Accident year	Development year						
	0	1	... k	... $n-i$	... $n-1$	n	
0	$Z_{0,0}$	$Z_{0,1}$	... $Z_{0,k}$	... $Z_{0,n-i}$	... $Z_{0,n-1}$	$Z_{0,n}$	
1	$Z_{1,0}$	$Z_{1,1}$	... $Z_{1,k}$	... $Z_{1,n-i}$	... $Z_{1,n-1}$	$Z_{1,n}$	
$\vdots$	$\vdots$	$\vdots$	$\vdots$	$\vdots$	$\vdots$	$\vdots$	
i	$Z_{i,0}$	$Z_{i,1}$	... $Z_{i,k}$	... $Z_{i,n-i}$	... $Z_{i,n-1}$	$Z_{i,n}$	
$\vdots$	$\vdots$	$\vdots$	$\vdots$	$\vdots$	$\vdots$	$\vdots$	
$n-k$	$Z_{n-k,0}$	$Z_{n-k,1}$	... $Z_{n-k,k}$	... $Z_{n-k,n-i}$	... $Z_{n-k,n-1}$	$Z_{n-k,n}$	
$\vdots$	$\vdots$	$\vdots$	$\vdots$	$\vdots$	$\vdots$	$\vdots$	
$n-1$	$Z_{n-1,0}$	$Z_{n-1,1}$	... $Z_{n-1,k}$	... $Z_{n-1,n-i}$	... $Z_{n-1,n-1}$	$Z_{n-1,n}$	
n	$Z_{n,0}$	$Z_{n,1}$	... $Z_{n,k}$	... $Z_{n,n-i}$	... $Z_{n,n-1}$	$Z_{n,n}$	

We assume that the incremental losses $Z_{i,k}$ are observable for $i + k \leq n$ and that they are non-observable for $i + k \geq n + 1$.

We also assume that

$$P[\{Z_{i,k} \in (0, \infty)\}] = 1$$

S. Kaulfuß · K.D. Schmidt (✉) · K. Thänert
Technische Universität Dresden, Dresden, Germany
e-mail: klaus.d.schmidt@tu-dresden.de

© Springer International Publishing Switzerland 2016
M. Radtke et al. (eds.), *Handbook on Loss Reserving*,
EAA Series, DOI 10.1007/978-3-319-30056-6_20

holds for all $i, k \in \{0, 1, \ldots, n\}$. This assumption is needed since we will use the logarithms of the incremental losses.

The assumption of strictly positive incremental losses is not fulfilled in every development triangle. For example, run-off triangles for *incurred losses* do not fulfill this assumption in general. This means that the lognormal loglinear model is not appropriate for such run-off triangles.

Let $m := (n+1)^2$ as well as $m_1 := (n+1)(n+2)/2$ and $m_2 := n(n+1)/2$. We represent the incremental losses $Z_{i,k}$ by a random vector

$$\mathbf{Z} = \begin{pmatrix} \mathbf{Z}_1 \\ \mathbf{Z}_2 \end{pmatrix}$$

such that $\mathbf{Z}_1$ consists of the m_1 observable incremental losses and $\mathbf{Z}_2$ consists of the m_2 non-observable incremental losses. Letting

$$\Sigma_{ik} := \mathrm{cov}[\ln(\mathbf{Z}_i), \ln(\mathbf{Z}_k)]$$

for $i, k \in \{1, 2\}$ and

$$\Sigma := \begin{pmatrix} \Sigma_{11} & \Sigma_{12} \\ \Sigma_{21} & \Sigma_{22} \end{pmatrix}$$

we obtain

$$\mathrm{var}[\ln(\mathbf{Z})] = \Sigma$$

In order to get a lognormal loglinear model for the incremental losses, we assume that

$$P_\mathbf{Z} = \mathbf{LN}(E[\ln(\mathbf{Z})], \Sigma)$$

and that there exist an unknown parameter $\beta \in \mathbb{R}^s$ with $s \leq m$ and known vectors $\mathbf{g}_{i,k} \in \mathbb{R}^s$ such that the identity

$$E[\ln(Z_{i,k})] = \mathbf{g}_{i,k}' \beta$$

holds for all $i, k \in \{0, 1, \ldots, n\}$. Arranging the row vectors $\mathbf{g}_{i,k}'$ in the same way as the incremental losses yields a matrix

$$\mathbf{A} = \begin{pmatrix} \mathbf{A}_1 \\ \mathbf{A}_2 \end{pmatrix}$$

such that

$$E\left[\begin{pmatrix} \ln(\mathbf{Z}_1) \\ \ln(\mathbf{Z}_2) \end{pmatrix}\right] = \begin{pmatrix} \mathbf{A}_1 \\ \mathbf{A}_2 \end{pmatrix} \beta$$

Since $P_{\mathbf{Z}} = \mathbf{LN}(E[\ln(\mathbf{Z})], \Sigma)$, the matrix Σ and hence Σ_{11} is positive definite. Thus, if $\operatorname{rang}(\mathbf{A}_1) = s$, then the assumptions of the *extended lognormal loglinear model* are fulfilled.

Lognormal Logadditive Model

The lognormal logadditive model consists of the following assumptions:

Lognormal Logadditive Model:

(i) *There exists a positive definite symmetric matrix Σ such that*

$$P_{\mathbf{Z}} = \mathbf{LN}(E[\ln(\mathbf{Z})], \Sigma)$$

(ii) *There exist unknown parameters μ as well as $\xi_1, \ldots, \xi_n$ and $\zeta_1, \ldots \zeta_n$ such that the identity*

$$E[\ln(Z_{i,k})] = \mu + (1 - \delta_{0,i})\,\xi_i + (1 - \delta_{0,k})\,\zeta_k$$

holds for all $i, k \in \{0, 1, \ldots, n\}$.

We assume in the following that the assumptions of the lognormal logadditive model are fulfilled and that the matrix Σ_{11} is known.

Then we have $s = 2n + 1$ and letting

$$\boldsymbol{\beta} := \begin{pmatrix} \mu \\ \xi_1 \\ \vdots \\ \xi_n \\ \zeta_1 \\ \vdots \\ \zeta_n \end{pmatrix}$$

and

$$\mathbf{g}_{i,k} := \mathbf{e}_1 + (1 - \delta_{0,i})\,\mathbf{e}_{1+i} + (1 - \delta_{0,k})\,\mathbf{e}_{1+n+k}$$

we obtain

$$E[\ln(Z_{i,k})] = \mathbf{g}'_{i,k}\boldsymbol{\beta}$$

Then the matrix $\mathbf{A}_1$ formed by the row vectors $\mathbf{g}'_{i,k}$ with $i + k \leq n$ fulfills the condition $\operatorname{rang}(\mathbf{A}_1) = s$. Therefore, the lognormal logadditive model is an *extended lognormal loglinear model*.

For all $i, k \in \{0, 1, \ldots, n\}$ there exists a unit vector $\mathbf{e}_{i,k} \in \mathbb{R}^m$ such that

$$Z_{i,k} = \mathbf{e}'_{i,k}\mathbf{Z}$$

and

$$\mathbf{g}'_{i,k} = \mathbf{e}'_{i,k}\mathbf{A}$$

This yields

$$
\begin{aligned}
E[Z_{i,k}] &= \exp\left(\mathbf{g}'_{i,k}\boldsymbol{\beta} + \frac{1}{2}\mathbf{e}'_{i,k}\Sigma\mathbf{e}_{i,k}\right) \\
&= E\left[\exp\left(\mathbf{g}'_{i,k}\boldsymbol{\beta}^{\mathrm{GM}} + \frac{1}{2}\mathbf{e}'_{i,k}\Sigma\mathbf{e}_{i,k} - \frac{1}{2}\mathbf{g}'_{i,k}(\mathbf{A}'_1\Sigma_{11}^{-1}\mathbf{A}_1)^{-1}\mathbf{g}_{i,k}\right)\right]
\end{aligned}
$$

where $\boldsymbol{\beta}^{\mathrm{GM}}$ is the *Gauss–Markov estimator* of $\boldsymbol{\beta}$. In this identity, the last expectation is that of an unbiased estimator $Z^*_{i,k}$ of the expectation of the incremental loss $Z_{i,k}$.

For $i + k \geq n + 1$ the incremental loss $Z_{i,k}$ is non-observable. Therefore, the unbiased estimator $Z^*_{i,k}$ of the expectation of the future incremental loss $Z_{i,k}$ can in this case also be understood as an unbiased *predictor* of the future incremental loss $Z_{i,k}$. Furthermore, the covariances

$$\mathrm{cov}[Z^*_{i,k} - Z_{i,k}, Z^*_{j,l} - Z_{j,l}]$$

of the prediction errors of the predictors of the future incremental losses can be specified explicitly.

If the predictors $Z^*_{i,k}$ with $i + k \geq n + 1$ are represented by a random vector $\mathbf{Z}^*_2$, then $\mathbf{Z}^*_2$ is an unbiased predictor of $\mathbf{Z}_2$ and we obtain an explicit formula for the variance

$$\mathrm{var}[\mathbf{Z}^*_2 - \mathbf{Z}_2]$$

of the prediction error. For every vector $\mathbf{d} \in \mathbb{R}^{m_2}$, $\mathbf{d}'\mathbf{Z}^*_2$ is then an unbiased predictor of the linear combination $\mathbf{d}'\mathbf{Z}_2$ of the future incremental losses and the expected squared prediction error of this predictor satisfies

$$
\begin{aligned}
E[(\mathbf{d}'\mathbf{Z}^*_2 - \mathbf{d}'\mathbf{Z}_2)^2] &= \mathrm{var}[\mathbf{d}'\mathbf{Z}^*_2 - \mathbf{d}'\mathbf{Z}_2] \\
&= \mathrm{var}[\mathbf{d}'(\mathbf{Z}^*_2 - \mathbf{Z}_2)] \\
&= \mathbf{d}'\mathrm{var}[\mathbf{Z}^*_2 - \mathbf{Z}_2]\mathbf{d}
\end{aligned}
$$

In particular, since reserves are sums of future incremental losses, we thus obtain, for every reserve, an unbiased predictor and a formula for the prediction error.

Example. Let $n = 2$ and $\Sigma = \mathbf{I}$. Then we have $m_1 = 6$ and $m_2 = 3$ as well as $s = 5$ and we obtain

$$\boldsymbol{\beta} = \begin{pmatrix} \mu \\ \xi_1 \\ \xi_2 \\ \zeta_1 \\ \zeta_2 \end{pmatrix}$$

Letting

$$\mathbf{Z}_1 := \begin{pmatrix} Z_{0,0} \\ Z_{0,1} \\ Z_{0,2} \\ Z_{1,0} \\ Z_{1,1} \\ Z_{2,0} \end{pmatrix} \quad \text{and} \quad \mathbf{A}_1 := \begin{pmatrix} 1\,0\,0\,0\,0 \\ 1\,0\,0\,1\,0 \\ 1\,0\,0\,0\,1 \\ 1\,1\,0\,0\,0 \\ 1\,1\,0\,1\,0 \\ 1\,0\,1\,0\,0 \end{pmatrix}$$

as well as

$$\mathbf{Z}_2 := \begin{pmatrix} Z_{1,2} \\ Z_{2,1} \\ Z_{2,2} \end{pmatrix} \quad \text{and} \quad \mathbf{A}_2 := \begin{pmatrix} 1\,1\,0\,0\,1 \\ 1\,0\,1\,1\,0 \\ 1\,0\,1\,0\,1 \end{pmatrix}$$

we then obtain

$$E[\ln(\mathbf{Z}_1)] = \mathbf{A}_1\boldsymbol{\beta} \quad \text{and} \quad E[\ln(\mathbf{Z}_2)] = \mathbf{A}_2\boldsymbol{\beta}$$

This yields

$$\mathbf{A}_1'\mathbf{A}_1 = \begin{pmatrix} 6\,2\,1\,2\,1 \\ 2\,2\,0\,1\,0 \\ 1\,0\,1\,0\,0 \\ 2\,1\,0\,2\,0 \\ 1\,0\,0\,0\,1 \end{pmatrix}$$

and hence

$$(\mathbf{A}_1'\mathbf{A}_1)^{-1} = \frac{1}{4}\begin{pmatrix} 3 & -2 & -3 & -2 & -3 \\ -2 & 4 & 2 & 0 & 2 \\ -3 & 2 & 7 & 2 & 3 \\ -2 & 0 & 2 & 4 & 2 \\ -3 & 2 & 3 & 2 & 7 \end{pmatrix}$$

Since $\Sigma = \mathbf{I}$ it then follows that the Gauss–Markov estimator of β satisfies

$$
\beta^{\mathrm{GM}} = \frac{1}{4}
\begin{pmatrix}
3 & 1 & 0 & 1 & -1 & 0 \\
-2 & -2 & 0 & 2 & 2 & 0 \\
-3 & -1 & 0 & -1 & 1 & 4 \\
-2 & 2 & 0 & -2 & 2 & 0 \\
-3 & -1 & 4 & -1 & 1 & 0
\end{pmatrix}
\begin{pmatrix}
\ln(Z_{0,0}) \\
\ln(Z_{0,1}) \\
\ln(Z_{0,2}) \\
\ln(Z_{1,0}) \\
\ln(Z_{1,1}) \\
\ln(Z_{2,0})
\end{pmatrix}
$$

We now determine the unbiased predictor $Z_{1,2}^{*}$ of $Z_{1,2}$. From

$$
\mathbf{g}_{1,2}' = \begin{pmatrix} 1 & 1 & 0 & 0 & 1 \end{pmatrix}
$$

we obtain

$$
\mathbf{g}_{1,2}'\beta^{\mathrm{GM}} = \ln\left(Z_{0,0}^{-1/2}\, Z_{0,1}^{-1/2}\, Z_{0,2}\, Z_{1,0}^{1/2}\, Z_{1,1}^{1/2} \right)
$$

and

$$
\mathbf{g}_{1,2}'(\mathbf{A}_1'\mathbf{A}_1)^{-1}\mathbf{g}_{1,2} = 2
$$

and hence

$$
Z_{1,2}^{*} = \exp\left(\mathbf{g}_{1,2}'\beta^{\mathrm{GM}} + \frac{1}{2}\mathbf{e}_{1,2}'\mathbf{e}_{1,2} - \frac{1}{2}\mathbf{g}_{1,2}'(\mathbf{A}_1'\mathbf{A}_1)^{-1}\mathbf{g}_{1,2} \right)
$$

$$
= \exp(-1/2)\left(\frac{Z_{1,0}Z_{1,1}}{Z_{0,0}Z_{0,1}} \right)^{1/2} Z_{0,2}
$$

Notes

Keywords: Linear Models (Basics), Linear Models (Loss Reserving), Lognormal Loglinear Model (Basics).

References: Doray [1996], Ludwig, Schmeißer & Thänert [2009], Verrall [1991].

Loss Development Method

Anja Schnaus

Consider the run-off square of cumulative losses:

Accident year	Development year						
	0	1	... k	... $n-i$	... $n-1$	n	
0	$S_{0,0}$	$S_{0,1}$	... $S_{0,k}$	... $S_{0,n-i}$	... $S_{0,n-1}$	$S_{0,n}$
1	$S_{1,0}$	$S_{1,1}$	... $S_{1,k}$	... $S_{1,n-i}$	... $S_{1,n-1}$	$S_{1,n}$
$\vdots$	$\vdots$	$\vdots$	$\vdots$	$\vdots$	$\vdots$	$\vdots$
i	$S_{i,0}$	$S_{i,1}$	... $S_{i,k}$	... $S_{i,n-i}$	... $S_{i,n-1}$	$S_{i,n}$
$\vdots$	$\vdots$	$\vdots$	$\vdots$	$\vdots$	$\vdots$	$\vdots$
$n-k$	$S_{n-k,0}$	$S_{n-k,1}$	... $S_{n-k,k}$	... $S_{n-k,n-i}$	... $S_{n-k,n-1}$	$S_{n-k,n}$
$\vdots$	$\vdots$	$\vdots$	$\vdots$	$\vdots$	$\vdots$	$\vdots$
$n-1$	$S_{n-1,0}$	$S_{n-1,1}$	... $S_{n-1,k}$	... $S_{n-1,n-i}$	... $S_{n-1,n-1}$	$S_{n-1,n}$
n	$S_{n,0}$	$S_{n,1}$	... $S_{n,k}$	... $S_{n,n-i}$	... $S_{n,n-1}$	$S_{n,n}$

We assume that the cumulative losses $S_{i,k}$ are observable for $i + k \leq n$ and that they are non-observable for $i + k \geq n + 1$. For $i, k \in \{0, 1, \ldots, n\}$ we denote by

$$
Z_{i,k} := \begin{cases} S_{i,0} & \text{if } k = 0 \\ S_{i,k} - S_{i,k-1} & \text{else} \end{cases}
$$

the incremental loss from accident year i in development year k.

A. Schnaus (✉)
Kerpen, Germany
e-mail: schnaus@genre.com

© Springer International Publishing Switzerland 2016
M. Radtke et al. (eds.), *Handbook on Loss Reserving*,
EAA Series, DOI 10.1007/978-3-319-30056-6_21

The loss development method is a prediction method based on the development pattern for quotas:

Development Pattern for Quotas: *There exist parameters $\gamma_0, \gamma_1, \ldots, \gamma_n$ with $\gamma_n = 1$ such that the identity*

$$\frac{E[S_{i,k}]}{E[S_{i,n}]} = \gamma_k$$

holds for all $k \in \{0, 1, \ldots, n\}$ and for all $i \in \{0, 1, \ldots, n\}$.

In this article we assume that a development pattern for quotas exists. Then the parameters $\vartheta_0, \vartheta_1, \ldots, \vartheta_n$ with

$$\vartheta_k := \begin{cases} \gamma_0 & \text{if } k = 0 \\ \gamma_k - \gamma_{k-1} & \text{else} \end{cases}$$

form a *development pattern for incremental quotas*. In particular, we are in the situation of a *multiplicative model*.

The loss development method predicts the future cumulative losses $S_{i,k}$ on the basis of the current losses $S_{i,n-i}$ using *a priori estimators*

$$\widehat{\gamma}_0, \widehat{\gamma}_1, \ldots, \widehat{\gamma}_n$$

with $\widehat{\gamma}_n := 1$ for the *quotas* $\gamma_0, \gamma_1, \ldots, \gamma_n$. The a priori estimators can be based

- exclusively on external information (*benchmarks*), which is not contained in the run-off triangle under consideration (for example on market statistics, on similar portfolios or on pricing assumptions), or
- exclusively on internal information, which is completely contained in the run-off triangle under consideration (for example on weighted means of the observable *individual development factors* $S_{i,k}/S_{i,k-1}$ as estimators of the *development factors* $\varphi_k := \gamma_k/\gamma_{k-1}$), or
- on a combination of external and internal information.

The a priori estimators are not subject to any limitation regarding the used information and are assumed as given.

The loss development method predicts, for every accident year i and every development year k such that $i + k \geq n + 1$, the future cumulative loss $S_{i,k}$ by the *loss development predictor*

$$\widehat{S}_{i,k}^{\text{LD}} := \widehat{\gamma}_k \frac{S_{i,n-i}}{\widehat{\gamma}_{n-i}}$$

Thus, the current loss $S_{i,n-i}$ is scaled first to the level of the ultimate loss $S_{i,n}$ (by division by the quota $\widehat{\gamma}_{n-i}$) and then to the level of the cumulative loss $S_{i,k}$ (by subsequent multiplication with the quota $\widehat{\gamma}_k$).

In particular, one has

$$\widehat{S}_{i,n}^{\mathrm{LD}} = \frac{S_{i,n-i}}{\widehat{\gamma}_{n-i}}$$

and hence

$$\widehat{S}_{i,k}^{\mathrm{LD}} = \widehat{\gamma}_k \, \widehat{S}_{i,n}^{\mathrm{LD}}$$

The definition of the loss development predictors of the cumulative losses replicates the identity

$$E[S_{i,k}] = \frac{E[S_{i,k}]}{E[S_{i,n}]} \frac{E[S_{i,n}]}{E[S_{i,n-i}]} E[S_{i,n-i}] = \gamma_k \, \frac{1}{\gamma_{n-i}} \, E[S_{i,n-i}]$$

which results from the development pattern for quotas.

From the loss development predictors of the future cumulative losses one obtains first the *loss development predictors*

$$\widehat{Z}_{i,k}^{\mathrm{LD}} := \begin{cases} \widehat{S}_{i,n-i+1}^{\mathrm{LD}} - S_{i,n-i} & \text{if } i + k = n + 1 \\ \widehat{S}_{i,k}^{\mathrm{LD}} - \widehat{S}_{i,k-1}^{\mathrm{LD}} & \text{else} \end{cases}$$

of the future incremental losses $Z_{i,k}$ with $i, k \in \{1, \ldots, n\}$ and $i + k \geq n + 1$ and then the *loss development predictors*

$$\widehat{R}_i^{\mathrm{LD}} := \sum_{l=n-i+1}^{n} \widehat{Z}_{i,l}^{\mathrm{LD}}$$

$$\widehat{R}_{(c)}^{\mathrm{LD}} := \sum_{l=c-n}^{n} \widehat{Z}_{c-l,l}^{\mathrm{LD}}$$

$$\widehat{R}^{\mathrm{LD}} := \sum_{l=1}^{n} \sum_{j=n-l+1}^{n} \widehat{Z}_{j,l}^{\mathrm{LD}}$$

of the accident year reserves R_i, with $i \in \{1, \ldots, n\}$, of the calendar year reserves $R_{(c)}$, with $c \in \{n+1, \ldots, 2n\}$, and of the aggregate loss reserve R. These reserve predictors are also called *loss development reserves*.

Example A. Calculation of the loss development predictors of cumulative losses:

Accident	Development year k					
year i	0	1	2	3	4	5
0						3483
1					3844	3963
2				3977	4286	4419
3			3880	4656	5018	5173
4		3261	4447	5336	5751	5929
5	1889	3463	4723	5667	6108	6297
$\widehat{\gamma}_k$	0.30	0.55	0.75	0.90	0.97	1

Incremental losses:

Accident	Development year k					
year i	0	1	2	3	4	5
0						
1						119
2					309	133
3				776	362	155
4			1186	889	415	178
5		1574	1259	945	441	189
$\widehat{\vartheta}_k$	0.30	0.25	0.20	0.15	0.07	0.03

Reserves:

Accident year i	Reserve R_i
1	119
2	442
3	1293
4	2668
5	4408
sum	8930

Calendar year c	Reserve $R_{(c)}$
6	3964
7	2643
8	1515
9	619
10	189
sum	8930

The estimators of the development pattern for incremental quotas are not required for the loss development method and serve only for comparison with other methods.

Example B. In this example the cumulative loss $S_{4,1}$ is increased by 1000:

Accident	Development year k					
year i	0	1	2	3	4	5
0						3483
1					3844	3963
2				3977	4286	4419
3			3880	4656	5018	5173
4		4261	5810	6973	7515	7747
5	1889	3463	4723	5667	6108	6297
$\widehat{\gamma}_k$	0.30	0.55	0.75	0.90	0.97	1

Incremental losses:

Accident	Development year k					
year i	0	1	2	3	4	5
0						
1						119
2					309	133
3				776	362	155
4			1549	1162	542	232
5		1574	1259	945	441	189
$\widehat{\vartheta}_k$	0.30	0.25	0.20	0.15	0.07	0.03

Reserves:

Accident year i	Reserve R_i
1	119
2	442
3	1293
4	3486
5	4408
sum	9748

Calendar year c	Reserve $R_{(c)}$
6	4328
7	2916
8	1642
9	673
10	189
sum	9748

The outlier $S_{4,1}$ affects the predictors of all incremental losses and of the reserve of accident year 4, and it also affects the predictors of the calendar year reserves of the calendar years 6 to 9.

Due to the multiplicative structure of the loss development predictors of cumulative losses, the current losses generate a leverage effect on the predictors of all future cumulative or incremental losses of the same accident year.

For the comparison with other methods it is useful to compute these examples with

- $\widehat{\gamma}_k := \gamma_k^{\mathrm{AD}}$ (*additive quotas*)
- $\widehat{\gamma}_k := \gamma_k^{\mathrm{CL}}$ (*chain ladder quotas*)
- $\widehat{\gamma}_k := \gamma_k^{\mathrm{PA}}$ (*Panning quotas*)

instead of the a priori estimators used here.

Bornhuetter–Ferguson Principle

Letting

$$\alpha_i^{\mathrm{LD}} := \frac{S_{i,n-i}}{\widehat{\gamma}_{n-i}}$$

the loss development predictors of the future cumulative losses can be represented
in the form

$$S_{i,k}^{LD} = S_{i,n-i} + \left(\widehat{\gamma}_k - \widehat{\gamma}_{n-i}\right) \widehat{\alpha}_i^{LD}$$

Therefore, the loss development method is subject to the *Bornhuetter–Ferguson
principle*.

Furthermore, in the case $\widehat{\gamma}_k := \gamma_k^{CL}$ the loss development method is equivalent to
the *chain ladder method*.

Remark

The loss development method takes a middle position between the *chain ladder
method* and the *Bornhuetter–Ferguson method*:

- The chain ladder method estimates the quotas by the chain ladder quotas and hence
  exclusively on the basis of the run-off triangle.
- The Bornhuetter–Ferguson method uses not only a priori estimators of the quotas,
  but also a priori estimators of the expected ultimate losses.

Furthermore, the predictors of the iterated Bornhuetter–Ferguson method converge
towards the predictors of the loss development method.

Notes

Keywords: Bornhuetter–Ferguson Method, Bornhuetter–Ferguson Principle, Chain
Ladder Method (Basics), Development Patterns (Basics), Development Patterns
(Estimation), Grossing Up Method, Multiplicative Models, Run-Off Triangles.

References: Bornhuetter & Ferguson [1972], Pierson [1994], Schmidt [2009, 2012],
Schmidt & Zocher [2008].

Loss Ratios

Michael Radtke and Anja Schnaus

Consider the run-off square of cumulative losses:

Accident year	Development year							
	0	1	... k	... $n-i$		... $n-1$	n	
0	$S_{0,0}$	$S_{0,1}$	... $S_{0,k}$	... $S_{0,n-i}$		... $S_{0,n-1}$	$S_{0,n}$	
1	$S_{1,0}$	$S_{1,1}$	... $S_{1,k}$	... $S_{1,n-i}$		... $S_{1,n-1}$	$S_{1,n}$	
$\vdots$	$\vdots$	$\vdots$	$\vdots$	$\vdots$		$\vdots$	$\vdots$	
i	$S_{i,0}$	$S_{i,1}$	... $S_{i,k}$	... $S_{i,n-i}$		... $S_{i,n-1}$	$S_{i,n}$	
$\vdots$	$\vdots$	$\vdots$	$\vdots$	$\vdots$		$\vdots$	$\vdots$	
$n-k$	$S_{n-k,0}$	$S_{n-k,1}$	... $S_{n-k,k}$	... $S_{n-k,n-i}$		... $S_{n-k,n-1}$	$S_{n-k,n}$	
$\vdots$	$\vdots$	$\vdots$	$\vdots$	$\vdots$		$\vdots$	$\vdots$	
$n-1$	$S_{n-1,0}$	$S_{n-1,1}$	... $S_{n-1,k}$	... $S_{n-1,n-i}$		... $S_{n-1,n-1}$	$S_{n-1,n}$	
n	$S_{n,0}$	$S_{n,1}$	... $S_{n,k}$	... $S_{n,n-i}$		... $S_{n,n-1}$	$S_{n,n}$	

We assume that the cumulative losses $S_{i,k}$ are observable for $i + k \leq n$ and that they are non-observable for $i + k \geq n + 1$.

In this article we consider reserving methods that use *a priori estimators*

$$\widehat{\alpha}_0, \widehat{\alpha}_1, \ldots, \widehat{\alpha}_n$$

of the expected ultimate losses $\alpha_0, \alpha_1, \ldots, \alpha_n$. Important representatives of this class of reserving methods are the *Bornhuetter–Ferguson method* and the *Cape Cod method*.

M. Radtke
Fachhochschule Dortmund, Dortmund, Germany
e-mail: michael.radtke@towerswatson.com

A. Schnaus (✉)
Kerpen, Germany
e-mail: schnaus@genre.com

© Springer International Publishing Switzerland 2016
M. Radtke et al. (eds.), *Handbook on Loss Reserving*,
EAA Series, DOI 10.1007/978-3-319-30056-6_22

The selection of the a priori estimators of the expected ultimate losses is aggravated by the fact that it is impossible to infer from the observable run-off triangle of cumulative losses whether an increase from one accident year to the next is caused, for example, by *large losses*, by underwriting new treaties or by other systematic effects.

In practice, a priori estimators of the expected ultimate losses are usually not chosen directly, but rather by starting with a priori estimators of the *expected ultimate loss ratios* relating to a *volume measure* (for example accounted, written or earned *premiums*) such that the cumulative losses are normalized by a volume measure.

For this purpose we consider the multiplicative model for ultimate loss ratios:

Multiplicative Model for Ultimate Loss Ratios: *There exist known volume measures* $v_0, v_1, \ldots, v_n$ *as well as unknown parameters* $\kappa_0, \kappa_1, \ldots, \kappa_n$ *and* $\gamma_0, \gamma_1, \ldots, \gamma_n$ *with* $\gamma_n = 1$ *such that the identity*

$$E\left[\frac{S_{i,k}}{v_i \gamma_k}\right] = \kappa_i$$

holds for every $i \in \{0, 1, \ldots, n\}$ *and for all* $k \in \{0, 1, \ldots, n\}$.

The multiplicative model for ultimate loss ratios is a *multiplicative model for cumulative losses* since it yields the identities

$$E[S_{i,k}] = (v_i \kappa_i)\, \gamma_k$$

Moreover, the parameters $\gamma_0, \gamma_1, \ldots, \gamma_n$ form a *development pattern for quotas* and it follows from the identity

$$\kappa_i = E\left[\frac{S_{i,n}}{v_i}\right]$$

that the parameters $\kappa_0, \kappa_1, \ldots, \kappa_n$ are the *expected ultimate loss ratios*.

In this article we assume that the multiplicative model for ultimate loss ratios is given.

For the estimation of the expected future cumulative losses $E[S_{i,k}]$ we use *a priori estimators*

$$\widehat{\gamma}_0, \widehat{\gamma}_1, \ldots, \widehat{\gamma}_n$$

with $\widehat{\gamma}_n := 1$ of the *quotas* $\gamma_0, \gamma_1, \ldots, \gamma_n$ and *a priori estimators*

$$\widehat{\kappa}_0, \widehat{\kappa}_1, \ldots, \widehat{\kappa}_n$$

of the expected ultimate loss ratios $\kappa_0, \kappa_1, \ldots, \kappa_n$. Furthermore we use

$$\widehat{S}_{i,k} := v_i \widehat{\kappa}_i \widehat{\gamma}_k$$

as an estimator of the expected cumulative loss $E[S_{i,k}]$. Then we obtain, in particular,

$$\widehat{\alpha}_i := v_i \widehat{\kappa}_i$$

as an *a priori estimators* of the expected ultimate loss $E[S_{i,n}]$.

This method for the estimation of the expected future cumulative losses on the basis of estimators of the expected ultimate loss ratios is highly sensitive with regard to the used volume measures. In the case where the volume measures are premiums, the estimators depend on the adequateness of pricing and the accounting and earning modalities of the premiums.

Estimation of the Expected Ultimate Loss Ratios

The *run-off triangle* of the *individual ultimate loss ratios*

$$\widehat{\kappa}_{i,k} := \frac{S_{i,k}}{v_i \widehat{\gamma}_k}$$

with $i, k \in \{0, 1, \ldots, n\}$ and $i + k \leq n$ is of particular interest since it supports the localization and the quantification of absent losses or *large losses*. These *outliers* can be taken into account in the selection of the estimators of the expected ultimate loss ratios.

Due to the assumptions of the multiplicative model for ultimate loss ratios, the identity

$$\kappa_i = E\left[\frac{S_{i,k}}{v_i \gamma_k}\right]$$

holds for every accident year $i \in \{0, 1, \ldots, n\}$ and for every development year $k \in \{0, 1, \ldots, n - i\}$. Hence it seems obvious to estimate the expected ultimate loss ratio κ_i by a weighted mean of the observable individual ultimate loss ratios $\widehat{\kappa}_{i,k}$ of accident year i.

Examples of such weighted means are

- the arithmetic mean

$$\widehat{\kappa}_i' := \frac{1}{n - i + 1} \sum_{l=0}^{n-i} \widehat{\kappa}_{i,l}$$

of all observable individual ultimate loss ratios of accident year i,

- the arithmetic mean

$$\widehat{\kappa}_i'' := \frac{1}{\min\{n-i+1, 3\}} \sum_{l=\max\{n-i-2,0\}}^{n-i} \widehat{\kappa}_{i,l}$$

of the observable individual ultimate loss ratios of the (at most) last three development years and
- the current individual ultimate loss ratio

$$\widehat{\kappa}_i''' := \widehat{\kappa}_{i,n-i} = \frac{S_{i,n-i}}{v_i \widehat{\gamma}_{n-i}} = \frac{S_{i,n}^{LD}}{v_i}$$

The last formula shows that the use of the current individual ultimate loss ratio corresponds to the *loss development method*.

The use of shortened arithmetic means like $\widehat{\kappa}_i''$ or $\widehat{\kappa}_i'''$ can be reasonable in the case of structural breaks, hence in the case of a changing development from one calendar year to the next.

In the case of large losses or absent losses it can be reasonable not to use the estimators considered before, which are specific for different accident years, but rather their weighted means of the form

$$\widehat{\kappa}^\circ := \sum_{j=0}^{n} \frac{v_j}{\sum_{h=0}^{n} v_h} \widehat{\kappa}_j^\circ$$

or

$$\widetilde{\kappa}^\circ := \sum_{j=0}^{n} \frac{v_j \widehat{\gamma}_{n-j}}{\sum_{h=0}^{n} v_h \widehat{\gamma}_{n-h}} \widehat{\kappa}_j^\circ$$

with $\circ \in \{', '', '''\}$ and weights that are proportional to either the volume measures (premiums) v_i or the *burnt premiums* $v_i \widehat{\gamma}_{n-i}$. Since

$$\widetilde{\kappa}''' = \sum_{j=0}^{n} \frac{v_j \widehat{\gamma}_{n-j}}{\sum_{h=0}^{n} v_h \widehat{\gamma}_{n-h}} \widehat{\kappa}_j''' = \sum_{j=0}^{n} \frac{v_j \widehat{\gamma}_{n-j}}{\sum_{h=0}^{n} v_h \widehat{\gamma}_{n-h}} \frac{S_{i,n-i}}{v_i \widehat{\gamma}_{n-i}} = \frac{\sum_{j=0}^{n} S_{j,n-j}}{\sum_{j=0}^{n} v_j \widehat{\gamma}_{n-j}}$$

we see that $\widetilde{\kappa}'''$ is just the *Cape Cod ultimate loss ratio* κ^{CC}.

Example A. Consider the run-off triangle of cumulative losses with volume measures and a priori estimators of the quotas

Accident	Development year k						Volume
year i	0	1	2	3	4	5	v_i
0	1001	1855	2423	2988	3335	3483	4025
1	1113	2103	2774	3422	3844		4456
2	1265	2433	3233	3977			5315
3	1490	2873	3880				5986
4	1725	3261					6939
5	1889						8158
$\widehat{\gamma}_k$	0.30	0.55	0.75	0.90	0.97	1	

The following table contains the realizations of the observable individual ultimate loss ratios and of the derived estimators:

Accident	Development year k						Estimators		
year i	0	1	2	3	4	5	$\widehat{\kappa}_i'$	$\widehat{\kappa}_i''$	$\widehat{\kappa}_i'''$
0	0.829	0.838	0.803	0.825	0.854	0.865	0.836	0.848	0.865
1	0.833	0.858	0.830	0.853	0.889		0.853	0.858	0.889
2	0.793	0.832	0.811	0.831			0.817	0.825	0.831
3	0.830	0.873	0.864				0.856	0.856	0.864
4	0.829	0.854					0.842	0.842	0.854
5	0.772						0.772	0.772	0.772
$\widehat{\kappa}', \widehat{\kappa}'', \widehat{\kappa}'''$							0.825	0.828	0.839
$\widetilde{\kappa}', \widetilde{\kappa}'', \widetilde{\kappa}'''$							0.833	0.838	0.851

The individual ultimate loss ratios of accident years 2 and 5 in development year 0 are relatively low and the Cape Cod ultimate loss ratio is relatively high.

Example B. In this example the cumulative loss $S_{4,1}$ is increased by 1000:

Accident	Development year k						Volume
year i	0	1	2	3	4	5	v_i
0	1001	1855	2423	2988	3335	3483	4025
1	1113	2103	2774	3422	3844		4456
2	1265	2433	3233	3977			5315
3	1490	2873	3880				5986
4	1725	4261					6939
5	1889						8158
$\widehat{\gamma}_k$	0.30	0.55	0.75	0.90	0.97	1	

The following table contains the realizations of the observable individual ultimate loss ratios and of the derived estimators:

Accident	Development year k						Estimators		
year i	0	1	2	3	4	5	$\widehat{\kappa}_i'$	$\widehat{\kappa}_i''$	$\widehat{\kappa}_i'''$
0	0.829	0.838	0.803	0.825	0.854	0.865	0.836	0.848	0.865
1	0.833	0.858	0.830	0.853	0.889		0.853	0.858	0.889
2	0.793	0.832	0.811	0.831			0.817	0.825	0.831
3	0.830	0.873	0.864				0.856	0.856	0.864
4	0.829	1.116					0.973	0.973	1.116
5	0.772						0.772	0.772	0.772
$\widehat{\kappa}', \widehat{\kappa}'', \widehat{\kappa}'''$							0.851	0.854	0.891
$\widetilde{\kappa}', \widetilde{\kappa}'', \widetilde{\kappa}'''$							0.854	0.859	0.893

The outlier $S_{4,1}$ is reflected in the individual ultimate loss ratio $\widehat{\kappa}_{4,1}$, hence in the weighted means of accident year 4, and thus in the means over the accident years.

Remarks

Multiplication of arbitrary estimators of the expected ultimate loss ratios with the volume measures provides a priori estimators of the expected ultimate losses, which are required in the *Bornhuetter–Ferguson method*.

The methods considered here to estimate the expected ultimate loss ratios from the run-off triangle were developed by Pierson (1994) in order to obtain robust a priori estimators of the expected ultimate losses.

Sometimes it is necessary to use estimators of the expected ultimate loss ratios that are exclusively based on external information not included in the run-off triangle under consideration (for example on market statistics, on similar portfolios or on pricing assumptions). In this case, the corresponding a priori estimators of the expected ultimate losses can be adapted to the run-off triangle via the *Benktander–Hovinen method* or, more generally, by the *iterated Bornhuetter–Ferguson method*.

Notes

Keywords: Bornhuetter–Ferguson Method, Bornhuetter–Ferguson Principle, Cape Cod Method, Chain Ladder Method (Basics), Development Pattern (Basics), Loss Development Method, Multiplicative Models, Run-Off Triangles.

References: Bornhuetter & Ferguson [1972], Pierson [1994].

Marginal Sum Method

Klaus D. Schmidt and Angela Wünsche

Consider the run-off square of incremental losses:

Accident year	Development year						
	0	1	... k	... $n-i$	... $n-1$	n
0	$Z_{0,0}$	$Z_{0,1}$	... $Z_{0,k}$	... $Z_{0,n-i}$	... $Z_{0,n-1}$	$Z_{0,n}$
1	$Z_{1,0}$	$Z_{1,1}$	... $Z_{1,k}$	... $Z_{1,n-i}$	... $Z_{1,n-1}$	$Z_{1,n}$
$\vdots$	$\vdots$	$\vdots$	$\vdots$	$\vdots$	$\vdots$	$\vdots$
i	$Z_{i,0}$	$Z_{i,1}$	... $Z_{i,k}$	... $Z_{i,n-i}$	... $Z_{i,n-1}$	$Z_{i,n}$
$\vdots$	$\vdots$	$\vdots$	$\vdots$	$\vdots$	$\vdots$	$\vdots$
$n-k$	$Z_{n-k,0}$	$Z_{n-k,1}$	... $Z_{n-k,k}$	... $Z_{n-k,n-i}$	... $Z_{n-k,n-1}$	$Z_{n-k,n}$
$\vdots$	$\vdots$	$\vdots$	$\vdots$	$\vdots$	$\vdots$	$\vdots$
$n-1$	$Z_{n-1,0}$	$Z_{n-1,1}$	... $Z_{n-1,k}$	... $Z_{n-1,n-i}$	... $Z_{n-1,n-1}$	$Z_{n-1,n}$
n	$Z_{n,0}$	$Z_{n,1}$	... $Z_{n,k}$	... $Z_{n,n-i}$	... $Z_{n,n-1}$	$Z_{n,n}$

We assume that the incremental losses $Z_{i,k}$ are observable for $i + k \leq n$ and that they are non-observable for $i + k \geq n + 1$. For $i \in \{0, 1, \ldots, n\}$ we denote by

$$S_{i,n} := \sum_{l=0}^{n} Z_{i,l}$$

the ultimate loss of accident year i.

The marginal sum method is based on the multiplicative model for incremental losses:

K.D. Schmidt (✉) · A. Wünsche
Technische Universität Dresden, Dresden, Germany
e-mail: klaus.d.schmidt@tu-dresden.de

© Springer International Publishing Switzerland 2016
M. Radtke et al. (eds.), *Handbook on Loss Reserving*,
EAA Series, DOI 10.1007/978-3-319-30056-6_23

Multiplicative Model for Incremental Losses: *There exist parameters $\alpha_0, \alpha_1, \ldots, \alpha_n$ and $\vartheta_0, \vartheta_1, \ldots, \vartheta_n$ with $\sum_{l=0}^{n} \vartheta_l = 1$ such that the identity*

$$E[Z_{i,k}] = \alpha_i \vartheta_k$$

holds for all $i, k \in \{0, 1, \ldots, n\}$.

In this article we assume that the assumptions of the multiplicative model for incremental losses are fulfilled. Then summation over all development years yields

$$E[S_{i,n}] = \alpha_i$$

Therefore, the parameters $\alpha_0, \alpha_1, \ldots, \alpha_n$ are the expected ultimate losses of the accident years, and it then follows that the parameters $\vartheta_0, \vartheta_1, \ldots, \vartheta_n$ form a *development pattern for incremental quotas*.

Estimation of the Parameters

A natural way to estimate the parameters is to use the *marginal sum method* in which the *marginal sum estimators*

$$\alpha_0^{MS}, \alpha_1^{MS}, \ldots, \alpha_n^{MS} \quad \text{and} \quad \vartheta_0^{MS}, \vartheta_1^{MS}, \ldots, \vartheta_n^{MS}$$

are determined as a solution of the *marginal sum equations*

$$\sum_{l=0}^{n-i} \widehat{\alpha_i} \widehat{\vartheta_l} = \sum_{l=0}^{n-i} Z_{i,l}$$

with $i \in \{0, 1, \ldots, n\}$ and

$$\sum_{j=0}^{n-k} \widehat{\alpha_j} \widehat{\vartheta_k} = \sum_{j=0}^{n-k} Z_{j,k}$$

with $k \in \{0, 1, \ldots, n\}$ under the constraint

$$\sum_{l=0}^{n} \widehat{\vartheta_l} = 1$$

The marginal sum equations replicate the identities

$$\sum_{l=0}^{n-i} \alpha_i \vartheta_l = \sum_{l=0}^{n-i} E[Z_{i,l}]$$

and

$$\sum_{j=0}^{n-k} \alpha_j \vartheta_k = \sum_{j=0}^{n-k} E[Z_{j,k}]$$

which result from the multiplicative model for incremental losses, and the constraint corresponds to the condition

$$\sum_{l=0}^{n} \vartheta_l = 1$$

This indicates that the marginal sum method is indeed a very natural method to estimate the parameters of the multiplicative model for incremental losses.

Since the development pattern for incremental quotas is equivalent to the development pattern for factors, the multiplicative model for incremental losses also provides the basis for the *chain ladder method*. The following result is quite remarkable:

Theorem. *Marginal sum estimators exist and are unique and they satisfy*

$$\widehat{\alpha}_i^{\mathrm{MS}} = S_{i,n-i} \prod_{l=n-i+1}^{n} \varphi_l^{\mathrm{CL}}$$

for every accident year $i \in \{0, 1, \ldots, n\}$ and

$$\vartheta_k^{\mathrm{MS}} = \begin{cases} \prod_{l=1}^{n} \dfrac{1}{\varphi_l^{\mathrm{CL}}} & if\ k = 0 \\[2ex] \prod_{l=k+1}^{n} \dfrac{1}{\varphi_l^{\mathrm{CL}}} - \prod_{l=k}^{n} \dfrac{1}{\varphi_l^{\mathrm{CL}}} & else \end{cases}$$

for every development year $k \in \{0, 1, \ldots, n\}$, with the chain ladder factors

$$\varphi_k^{\mathrm{CL}} := \frac{\sum_{j=0}^{n-k} S_{j,k}}{\sum_{j=0}^{n-k} S_{j,k-1}}$$

In particular, for every accident year $i \in \{1, \ldots, n\}$, the marginal sum estimator of the expected ultimate losses is identical with the chain ladder predictor of the ultimate losses.

The marginal sum method thus provides a simple justification of the *chain ladder method* by a general principle of estimation.

Remark

The marginal sum method has its origin in pricing, where two tariff criteria with a finite number of values are considered instead of accident years and development years.

Notes

Keywords: Chain Ladder Method (Basics), Development Patterns (Basics), Multiplicative Models.

References: Dietze, Riedrich & Schmidt [2006, 2011], Mack [1991, 2002], Schmidt [2009], Schmidt & Wünsche [1998].

Multinomial Model

Klaus Th. Hess, Klaus D. Schmidt and Angela Wünsche

Consider the run-off square of incremental losses:

Accident year	Development year					
	0	1	... k	... $n-i$	... $n-1$	n
0	$Z_{0,0}$	$Z_{0,1}$	... $Z_{0,k}$	... $Z_{0,n-i}$	... $Z_{0,n-1}$	$Z_{0,n}$
1	$Z_{1,0}$	$Z_{1,1}$	... $Z_{1,k}$	... $Z_{1,n-i}$	... $Z_{1,n-1}$	$Z_{1,n}$
$\vdots$	$\vdots$	$\vdots$	$\vdots$	$\vdots$	$\vdots$	$\vdots$
i	$Z_{i,0}$	$Z_{i,1}$	... $Z_{i,k}$	... $Z_{i,n-i}$	... $Z_{i,n-1}$	$Z_{i,n}$
$\vdots$	$\vdots$	$\vdots$	$\vdots$	$\vdots$	$\vdots$	$\vdots$
$n-k$	$Z_{n-k,0}$	$Z_{n-k,1}$	... $Z_{n-k,k}$	... $Z_{n-k,n-i}$	... $Z_{n-k,n-1}$	$Z_{n-k,n}$
$\vdots$	$\vdots$	$\vdots$	$\vdots$	$\vdots$	$\vdots$	$\vdots$
$n-1$	$Z_{n-1,0}$	$Z_{n-1,1}$	... $Z_{n-1,k}$	... $Z_{n-1,n-i}$	... $Z_{n-1,n-1}$	$Z_{n-1,n}$
n	$Z_{n,0}$	$Z_{n,1}$	... $Z_{n,k}$	... $Z_{n,n-i}$	... $Z_{n,n-1}$	$Z_{n,n}$

We assume that the incremental losses $Z_{i,k}$ are observable for $i + k \leq n$ and that they are non-observable for $i + k \geq n + 1$. For $i \in \{0, 1, \ldots, n\}$ we denote by

$$S_{i,n} := \sum_{l=0}^{n} Z_{i,l}$$

K.Th. Hess
Universität Rostock, Rostock, Germany
e-mail: klaus-thomas.hess@uni-rostock.de

K.D. Schmidt (✉) · A. Wünsche
Technische Universität Dresden, Dresden, Germany
e-mail: klaus.d.schmidt@tu-dresden.de

© Springer International Publishing Switzerland 2016
M. Radtke et al. (eds.), *Handbook on Loss Reserving*,
EAA Series, DOI 10.1007/978-3-319-30056-6_24

the ultimate loss of accident year i.

The multinomial model is a model for *claim numbers*.

Multinomial Model:

(i) *The incremental losses take only values in the set $\mathbb{N}_0$.*

(ii) *There exist parameters $\vartheta_0, \vartheta_1, \ldots, \vartheta_n \in (0, 1)$ with $\sum_{l=0}^{n} \vartheta_l = 1$ such that*

$$P_{Z_{i,0}, Z_{i,1}, \ldots, Z_{i,n} | S_{i,n}} = \mathbf{M}(S_{i,n}; \vartheta_0, \vartheta_1, \ldots, \vartheta_n)$$

holds for all $i \in \{0, 1, \ldots, n\}$.

In the multinomial model one has $P_{Z_{i,k} | S_{i,n}} = \mathbf{B}(S_{i,n}; \vartheta_k)$ and hence

$$E(Z_{i,k} | S_{i,n}) = S_{i,n} \, \vartheta_k$$

which can also be written as

$$E\left(\frac{Z_{i,k}}{S_{i,n}} \,\middle|\, S_{i,n}\right) = \vartheta_k$$

These identities present a refinement of the notion of a development pattern for incremental losses. Indeed, taking expectations yields

$$\frac{E[Z_{i,k}]}{E[S_{i,n}]} = \vartheta_k = E\left[\frac{Z_{i,k}}{S_{i,n}}\right]$$

Because of the left identity, the parameters $\vartheta_0, \vartheta_1, \ldots, \vartheta_n$ form a *development pattern for incremental quotas*, and it follows that the multinomial model is a *multiplicative model*. Moreover, because of the right identity, the expected individual incremental quotas $E[Z_{i,k}/S_{i,n}]$ are also independent of the accident year i. This shows that in the multinomial model the basic idea of a development pattern is formalized more precisely than it is in other models of loss reserving.

The previous identities also yield

$$\mathrm{cov}\left[\frac{Z_{i,k}}{S_{i,n}}, S_{i,n}\right] = 0$$

which means that the *individual incremental quota* $Z_{i,k}/S_{i,n}$ and the ultimate loss $S_{i,n}$ are uncorrelated in the multinomial model.

Justification of the Multinomial Model

The multinomial model can be justified by the following run-off model for claim numbers:

Run-Off Model for Claim Numbers:

(i) *The incremental losses take only values in the set $\mathbb{N}_0$.*

(ii) *For every accident year $i \in \{0, 1, \ldots, n\}$ there exists a sequence of random variables $\{K_{i,r}\}_{r \in \mathbb{N}}$ with values in $\{0, 1, \ldots, n\}$ which is independent and identically distributed and also independent of the ultimate loss $S_{i,n}$.*

(iii) *There exist parameters $\vartheta_0, \vartheta_1, \ldots, \vartheta_n \in (0, 1)$ with $\sum_{l=0}^{n} \vartheta_l = 1$ and*

$$P[\{K_{i,r} = k\}] = \vartheta_k$$

for all $i, k \in \{0, 1, \ldots, n\}$ and $r \in \mathbb{N}$.

In the run-off model for claim numbers

- $S_{i,n}$ is interpreted as the number of claims from accident year i and
- $K_{i,r}$ is interpreted as the development year in which claim r from accident year i is reported.

With this interpretation,

$$Z_{i,k} := \sum_{r=1}^{S_{i,n}} \chi_{\{K_{i,r}=k\}}$$

is the number of claims from accident year i which are reported or processed in development year k. In particular, every incremental loss $Z_{i,k}$ is modelled by a *collective model*. Furthermore, since

$$\sum_{l=0}^{n} Z_{i,l} = S_{i,n}$$

and

$$P_{Z_{i,0}, Z_{i,1}, \ldots, Z_{i,n} | S_{i,n}} = \mathbf{M}(S_{i,n}; \vartheta_0, \vartheta_1, \ldots, \vartheta_n)$$

the run-off model for claim numbers leads indeed to a multinomial model.

Marginal Sum Method

As the multinomial model is a multiplicative model, the *marginal sum method* can be used to estimate the expected ultimate losses $\alpha_0, \alpha_1, \ldots, \alpha_n$ and the development pattern $\vartheta_0, \vartheta_1, \ldots, \vartheta_n$.

Multinomial Model with Independent Accident Years

The assumptions of the multinomial model yield, for every accident year $i \subset \{0, 1, \ldots, n\}$ and for every family $\{z_{i,k}\}_{k \in \{0,1,\ldots,n\}} \subseteq \mathbb{N}_0$ and $s_{i,n} := \sum_{l=0}^{n} z_{i,l}$,

$$P\left[\bigcap_{l=0}^{n} \{Z_{i,l} = z_{i,l}\}\right] = P\left[\bigcap_{l=0}^{n} \{Z_{i,l} = z_{i,l}\} \,\middle|\, \{S_{i,n} = s_{i,n}\}\right] P[\{S_{i,n} = s_{i,n}\}]$$

$$= \left(\frac{s_{i,n}!}{\prod_{l=0}^{n} z_{i,l}!} \prod_{l=0}^{n} \vartheta_l^{z_{i,l}}\right) P[\{S_{i,n} = s_{i,n}\}]$$

Thus, for every accident year i, the joint distribution of the incremental losses is known except for the distribution of the ultimate loss $S_{i,n}$.

We consider now the multinomial model with independent accident years:

Multinomial Model with Independent Accident Years:

  (i) *The incremental losses take only values in the set* $\mathbb{N}_0$.
  (ii) *The family* $\{\{Z_{i,k}\}_{k \in \{0,1,\ldots,n\}}\}_{i \in \{0,1,\ldots,n\}}$ *is independent.*
  (iii) *There exist parameters* $\vartheta_0, \vartheta_1, \ldots, \vartheta_n \in (0, 1)$ *with* $\sum_{l=0}^{n} \vartheta_l = 1$ *such that*

$$P_{Z_{i,0}, Z_{i,1}, \ldots, Z_{i,n} | S_{i,n}} = \mathbf{M}(S_{i,n}; \vartheta_0; \vartheta_1, \ldots, \vartheta_n)$$

  holds for all $i \in \{0, 1, \ldots, n\}$.

Examples of a multinomial model with independent accident years are given by the *credibility model of Witting* and the *Poisson model*.

In the multinomial model with independent accident years one has

$$P\left[\bigcap_{j=0}^{n} \bigcap_{l=0}^{n} \{Z_{j,l} = z_{j,l}\}\right] = \prod_{j=0}^{n} P\left[\bigcap_{l=0}^{n} \{Z_{j,l} = z_{j,l}\}\right]$$

and then the identity for the joint distribution of the incremental losses from accident year i yields

$$P\left[\bigcap_{j=0}^{n} \bigcap_{l=0}^{n} \{Z_{j,l} = z_{j,l}\}\right] = \prod_{j=0}^{n}\left(\left(\frac{s_{j,n}!}{\prod_{l=0}^{n} z_{j,l}!} \prod_{l=0}^{n} \vartheta_l^{z_{j,l}}\right) P[\{S_{j,n} = s_{j,n}\}]\right)$$

with $s_{i,n} := \sum_{l=0}^{n} z_{i,l}$. Thus, the joint distribution of all incremental losses is known except for the distributions of the ultimate losses.

Maximum Likelihood Method

In order to determine completely, up to the parameters, the joint distribution of all incremental losses in the multinomial model with independent accident years, it is thus sufficient to select parametric classes of distributions for all ultimate losses. If this is done, then the joint distribution of the observable incremental losses is known except for the parameters, and the *maximum likelihood method* can be used to estimate the parameters. In doing so, the constraint

$$\sum_{l=0}^{n} \vartheta_l = 1$$

has to be taken into account.

We consider now three special cases which correspond to the three families of distributions belonging to the *Panjer class*:

Poisson Distributed Ultimate Losses

In the multinomial model with independent accident years, the assumption

$$P_{S_{i,n}} = \mathbf{Poi}(\alpha_i)$$

yields

$$P\left[\bigcap_{j=0}^{n}\bigcap_{l=0}^{n}\{Z_{j,l} = z_{j,l}\}\right] = \prod_{j=0}^{n}\prod_{l=0}^{n}\left(e^{-\alpha_j\vartheta_l}\,\frac{(\alpha_j\vartheta_l)^{z_{j,l}}}{z_{j,l}!}\right)$$

Thus, in this case the multinomial model becomes the *Poisson model* and it follows that the maximum likelihood estimators of the parameters are identical with the marginal sum estimators. The following result shows the particularity of the Poisson assumption:

Theorem. *In the multinomial model, the following assertions are equivalent:*

(a) *The accident years are independent with*

$$P_{S_{i,n}} = \mathbf{Poi}(\alpha_i)$$

(b) *The family of all incremental losses is independent with*

$$P_{Z_{i,k}} = \mathbf{Poi}(\alpha_i\vartheta_k)$$

(c) *The family of all incremental losses is independent.*

Therefore, the Poisson model is the only *multinomial model with independent incremental losses.*

Negative Binomial Distributed Ultimate Losses

In the multinomial model with independent accident years, the assumption

$$P_{S_{i,n}} = \mathbf{NB}(\beta_i, \eta_i)$$

yields

$$P\left[\bigcap_{j=0}^{n}\bigcap_{l=0}^{n}\{Z_{j,l} = z_{j,l}\}\right] = \prod_{j=0}^{n}\left(\frac{\Gamma(\beta_j + \sum_{l=0}^{n} z_{j,l})}{\Gamma(\beta_j)\prod_{l=0}^{n} z_{j,l}!}(1-\eta_j)^{\beta_j}\prod_{l=0}^{n}(\eta_j\vartheta_l)^{z_{j,l}}\right)$$

and summation gives

$$P\left[\bigcap_{j=0}^{n}\bigcap_{l=0}^{n-j}\{Z_{j,l} = z_{j,l}\}\right]$$

$$= \prod_{j=0}^{n}\left(\frac{\Gamma(\beta_j + \sum_{l=0}^{n-j} z_{j,l})}{\Gamma(\beta_j)\prod_{l=0}^{n-j} z_{j,l}!}\left(\frac{1-\eta_j}{1-\eta_j + \sum_{l=0}^{n-j}\eta_j\vartheta_l}\right)^{\beta_j}\prod_{l=0}^{n-j}\left(\frac{\eta_j\vartheta_l}{1-\eta_j + \sum_{l=0}^{n-j}\eta_j\vartheta_l}\right)^{z_{j,l}}\right)$$

so that the joint distribution of the observable incremental losses is known except for the parameters.

In contrast to the Poisson case, the expected ultimate losses α_i do not appear in the previous identity, but they can be introduced by a change of parameters: The assumption $P_{S_{i,n}} = \mathbf{NB}(\beta_i, \eta_i)$ yields

$$\alpha_i = E[S_{i,n}] = \beta_i\eta_i/(1-\eta_i)$$

and hence

$$\beta_i = \alpha_i(1-\eta_i)/\eta_i$$

Therefore, the joint distribution of the observable incremental losses can be represented in terms of the parameters $\alpha_0, \alpha_1, \ldots, \alpha_n$ as well as $\vartheta_0, \vartheta_1, \ldots, \vartheta_n$ and $\eta_0, \eta_1, \ldots, \eta_n$.

The maximum likelihood method can now be applied as in the Poisson case, taking into account that all parameters of the model have to be estimated although the parameters $\eta_0, \eta_1, \ldots, \eta_n$ are without interest. It turns out that also in the negative binomial case the maximum likelihood estimators of the expected ultimate losses and of the parameters of the development pattern are identical with the marginal sum estimators.

Binomial Distributed Ultimate Losses

In the multinomial model with independent accident years and with

$$P_{S_{i,n}} = \mathbf{B}(m_i; \eta_i)$$

one can proceed as in the negative binomial case and obtain similar results. The binomial case is of interest when in accident year i the portfolio consists of m_i independent risks and every risk suffers a loss with probability η_i.

Remark

The negative binomial distribution is a mixed Poisson distribution and is, like the Poisson distribution, a member of the family of *Hofmann distributions*. The results for negative binomial distributed ultimate losses can be generalized to Hofmann distributed ultimate losses.

Notes

Keywords: Chain Ladder Method (Basics), Collective Model, Credibility Models (Loss Reserving), Development Patterns (Basics), Marginal Sum Method, Multiplicative Models, Poisson Model, Simulation.

References: Schiegl [2002], Schmidt [2009], Schmidt & Wünsche [1998], Schmidt & Zocher [2005].

Multiplicative Models

Klaus D. Schmidt

Many models and methods of loss reserving are based on the idea that the expected incremental or cumulative losses can be represented as a product of two parameters such that one of the parameters depends only on the accident year while the other one depends only on the development year. In such a multiplicative structure it is reasonable to normalize either the parameters of the accident years or those of the development years in order to avoid too many parameters.

In this article we consider two multiplicative models in which the parameters of the development years are normalized and form a *development pattern*.

Multiplicative Model for Incremental Losses

Consider the run-off square of incremental losses:

Accident year	Development year						
	0	1	... k	... $n-i$	... $n-1$	n	
0	$Z_{0,0}$	$Z_{0,1}$	... $Z_{0,k}$	... $Z_{0,n-i}$	... $Z_{0,n-1}$	$Z_{0,n}$
1	$Z_{1,0}$	$Z_{1,1}$	... $Z_{1,k}$	... $Z_{1,n-i}$	... $Z_{1,n-1}$	$Z_{1,n}$
$\vdots$	$\vdots$	$\vdots$	$\vdots$	$\vdots$	$\vdots$	$\vdots$
i	$Z_{i,0}$	$Z_{i,1}$	... $Z_{i,k}$	... $Z_{i,n-i}$	... $Z_{i,n-1}$	$Z_{i,n}$
$\vdots$	$\vdots$	$\vdots$	$\vdots$	$\vdots$	$\vdots$	$\vdots$
$n-k$	$Z_{n-k,0}$	$Z_{n-k,1}$	... $Z_{n-k,k}$	... $Z_{n-k,n-i}$	... $Z_{n-k,n-1}$	$Z_{n-k,n}$
$\vdots$	$\vdots$	$\vdots$	$\vdots$	$\vdots$	$\vdots$	$\vdots$
$n-1$	$Z_{n-1,0}$	$Z_{n-1,1}$	... $Z_{n-1,k}$	... $Z_{n-1,n-i}$	... $Z_{n-1,n-1}$	$Z_{n-1,n}$
n	$Z_{n,0}$	$Z_{n,1}$	... $Z_{n,k}$	... $Z_{n,n-i}$	... $Z_{n,n-1}$	$Z_{n,n}$

K.D. Schmidt (✉)
Technische Universität Dresden, Dresden, Germany
e-mail: klaus.d.schmidt@tu-dresden.de

© Springer International Publishing Switzerland 2016
M. Radtke et al. (eds.), *Handbook on Loss Reserving*,
EAA Series, DOI 10.1007/978-3-319-30056-6_25

For $i, k \in \{0, 1, \ldots, n\}$ we denote by

$$S_{i,k} := \sum_{l=0}^{k} Z_{i,l}$$

the cumulative loss from accident year i in development year k.

In the multiplicative model for incremental losses, the sum of the parameters of the development years is normalized to 1:

Multiplicative Model for Incremental Losses: *There exist parameters $\alpha_0, \alpha_1, \ldots, \alpha_n$ and $\vartheta_0, \vartheta_1, \ldots, \vartheta_n$ with $\sum_{l=0}^{n} \vartheta_l = 1$ such that the identity*

$$E[Z_{i,k}] = \alpha_i \vartheta_k$$

holds for all $i, k \in \{0, 1, \ldots, n\}$.

In the multiplicative model for incremental losses one has

$$E[S_{i,n}] = \alpha_i$$

such that the parameters $\alpha_0, \alpha_1, \ldots, \alpha_n$ are the expected ultimate losses of the accident years and the parameters $\vartheta_0, \vartheta_1, \ldots, \vartheta_n$ form a *development pattern for incremental quotas*. Therefore, the multiplicative model for incremental losses provides the basis for every method of loss reserving which is based on a development pattern for incremental quotas, quotas or factors.

Multiplicative Model for Cumulative Losses

Consider the run-off square of cumulative losses:

Accident year	Development year						
	0	1	... k	... n−i	... n−1	n
0	$S_{0,0}$	$S_{0,1}$	... $S_{0,k}$	... $S_{0,n-i}$	... $S_{0,n-1}$	$S_{0,n}$
1	$S_{1,0}$	$S_{1,1}$	... $S_{1,k}$	... $S_{1,n-i}$	... $S_{1,n-1}$	$S_{1,n}$
⋮	⋮	⋮	⋮	⋮	⋮	⋮
i	$S_{i,0}$	$S_{i,1}$	... $S_{i,k}$	... $S_{i,n-i}$	... $S_{i,n-1}$	$S_{i,n}$
⋮	⋮	⋮	⋮	⋮	⋮	⋮
$n-k$	$S_{n-k,0}$	$S_{n-k,1}$	... $S_{n-k,k}$	... $S_{n-k,n-i}$	... $S_{n-k,n-1}$	$S_{n-k,n}$
⋮	⋮	⋮	⋮	⋮	⋮	⋮
$n-1$	$S_{n-1,0}$	$S_{n-1,1}$	... $S_{n-1,k}$	... $S_{n-1,n-i}$	... $S_{n-1,n-1}$	$S_{n-1,n}$
n	$S_{n,0}$	$S_{n,1}$	... $S_{n,k}$	... $S_{n,n-i}$	... $S_{n,n-1}$	$S_{n,n}$

In the multiplicative model for cumulative losses the parameter of the last development year is normalized to 1:

Multiplicative Model for Cumulative Losses: *There exist parameters $\alpha_0, \alpha_1, \ldots, \alpha_n$ and $\gamma_0, \gamma_1, \ldots, \gamma_n$ with $\gamma_n = 1$ such that the identity*

$$E[S_{i,k}] = \alpha_i \gamma_k$$

holds for all $i, k \in \{0, 1, \ldots, n\}$.

In the multiplicative model for cumulative losses one has

$$E[S_{i,n}] = \alpha_i$$

such that the parameters $\alpha_0, \alpha_1, \ldots, \alpha_n$ are the expected ultimate losses of the accident years and the parameters $\gamma_0, \gamma_1, \ldots, \gamma_n$ form a *development pattern for quotas*. Therefore, the multiplicative model for cumulative losses provides the basis for every method of loss reserving which is based on a development pattern for incremental quotas, quotas or factors.

Comparison

The two multiplicative models are equivalent since the respective development patterns are equivalent:

Theorem.

(1) *The multiplicative model for cumulative losses results from the multiplicative model for incremental losses by letting*

$$\gamma_k := \sum_{l=0}^{k} \vartheta_l$$

(2) *The multiplicative model for incremental losses results from the multiplicative model for cumulative losses by letting*

$$\vartheta_k := \begin{cases} \gamma_0 & \text{if } k = 0 \\ \gamma_k - \gamma_{k-1} & \text{else} \end{cases}$$

Assuming that one and hence each of the two multiplicative models exists means that an assumption is made on the joint distribution of all incremental or cumulative losses. Therefore, multiplicative models are stochastic models. They are particularly simple as they contain only assumptions on the structure of the expectations of incremental or cumulative losses.

Multiplicative Models and Development Patterns

Essentially, multiplicative models for incremental or cumulative losses are nothing else than development patterns for incremental quotas or quotas. The difference consists in the fact that development patterns focus on properties of development years while multiplicative models use assumptions on accident years and development years, which are almost symmetric.

Remarks

Multiplicative models with additional assumptions are given by the *credibility model of Mack*, the *credibility model of Witting* and the *credibility model of Hesselager and Witting* as well as in every *multinomial model* and hence, in particular, in the *Poisson model*.

To estimate the parameters of the multiplicative model for incremental losses one may use the *marginal sum method*. However, in the case of multiplicative models with additional assumptions, other methods taking into account these additional assumptions are preferable.

Multiplicative models have their origin in pricing where two tariff criteria with a finite number of values are considered instead of accident years and development years.

Notes

Keywords: Credibility Models (Loss Reserving), Development Patterns (Basics), Development Patterns (Estimation), Marginal Sum Method, Multinomial Model, Poisson Model.

References: Hess [2009], Mack [1991, 2002], Radtke [2008], Reich & Zeller [1988], Schmidt [2009], Schmidt & Wünsche [1998].

Multivariate Methods

Klaus Th. Hess, Alexander Ludwig, Carsten Pröhl, Klaus D. Schmidt and Mathias Zocher

Multivariate methods aim at simultaneous loss prediction for several lines of business. The advantage resulting from the use of multivariate methods instead of univariate methods is due to the fact that, for a particular line of business, multivariate methods use not only the information provided by this line of business but also that provided by the other lines of business under consideration. In particular, multivariate methods can take into account correlations between different lines of business.

We consider a portfolio consisting of M sub-portfolios (lines of business) and the corresponding run-off square of multivariate incremental losses:

Accident year	Development year						
	0	1	... k	... $n-i$	... $n-1$	n	
0	$\mathbf{Z}_{0,0}$	$\mathbf{Z}_{0,1}$	... $\mathbf{Z}_{0,k}$	... $\mathbf{Z}_{0,n-i}$	... $\mathbf{Z}_{0,n-1}$	$\mathbf{Z}_{0,n}$	
1	$\mathbf{Z}_{1,0}$	$\mathbf{Z}_{1,1}$	... $\mathbf{Z}_{1,k}$	... $\mathbf{Z}_{1,n-i}$	... $\mathbf{Z}_{1,n-1}$	$\mathbf{Z}_{1,n}$	
$\vdots$	$\vdots$	$\vdots$	$\vdots$	$\vdots$	$\vdots$	$\vdots$	
i	$\mathbf{Z}_{i,0}$	$\mathbf{Z}_{i,1}$	... $\mathbf{Z}_{i,k}$	... $\mathbf{Z}_{i,n-i}$	... $\mathbf{Z}_{i,n-1}$	$\mathbf{Z}_{i,n}$	
$\vdots$	$\vdots$	$\vdots$	$\vdots$	$\vdots$	$\vdots$	$\vdots$	
$n-k$	$\mathbf{Z}_{n-k,0}$	$\mathbf{Z}_{n-k,1}$	... $\mathbf{Z}_{n-k,k}$	... $\mathbf{Z}_{n-k,n-i}$	... $\mathbf{Z}_{n-k,n-1}$	$\mathbf{Z}_{n-k,n}$	
$\vdots$	$\vdots$	$\vdots$	$\vdots$	$\vdots$	$\vdots$	$\vdots$	
$n-1$	$\mathbf{Z}_{n-1,0}$	$\mathbf{Z}_{n-1,1}$	... $\mathbf{Z}_{n-1,k}$	... $\mathbf{Z}_{n-1,n-i}$	... $\mathbf{Z}_{n-1,n-1}$	$\mathbf{Z}_{n-1,n}$	
n	$\mathbf{Z}_{n,0}$	$\mathbf{Z}_{n,1}$	... $\mathbf{Z}_{n,k}$	... $\mathbf{Z}_{n,n-i}$	... $\mathbf{Z}_{n,n-1}$	$\mathbf{Z}_{n,n}$	

K.T. Hess
Universität Rostock, Rostock, Germany
e-mail: klaus-thomas.hess@uni-rostock.de

A. Ludwig · C. Pröhl · K.D. Schmidt (✉) · M. Zocher
Technische Universität Dresden, Dresden, Germany
e-mail: klaus.d.schmidt@tu-dresden.de

© Springer International Publishing Switzerland 2016
M. Radtke et al. (eds.), *Handbook on Loss Reserving*,
EAA Series, DOI 10.1007/978-3-319-30056-6_26

The multivariate incremental losses have the form

$$\mathbf{Z}_{i,k} = \begin{pmatrix} Z_{i,k}^{(1)} \\ \vdots \\ Z_{i,k}^{(M)} \end{pmatrix}$$

where $Z_{i,k}^{(m)}$ denotes the incremental loss of sub-portfolio $m \in \{1, \ldots, M\}$.

Summation of the multivariate incremental losses yields the multivariate cumulative losses

$$\mathbf{S}_{i,k} := \sum_{l=0}^{k} \mathbf{Z}_{i,l}$$

We assume that the (multivariate) incremental losses and hence also the (multivariate) cumulative losses are observable for $i + k \leq n$ and that they are non-observable for $i + k \geq n + 1$.

Multivariate Additive Method

The multivariate additive method is based on the *multivariate additive model*:

Multivariate Additive Model: *There exist known positive definite diagonal matrices* $\mathbf{V}_0, \mathbf{V}_1, \ldots, \mathbf{V}_n$ *as well as unknown parameters* $\boldsymbol{\zeta}_0, \boldsymbol{\zeta}_1, \ldots, \boldsymbol{\zeta}_n$ *and positive definite symmetric matrices* $\boldsymbol{\Sigma}_0, \boldsymbol{\Sigma}_1, \ldots, \boldsymbol{\Sigma}_n$ *such that the identities*

$$E[\mathbf{Z}_{i,k}] = \mathbf{V}_i \, \boldsymbol{\zeta}_k$$
$$\mathrm{cov}[\mathbf{Z}_{i,k}, \mathbf{Z}_{j,l}] = \mathbf{V}_i^{1/2} \boldsymbol{\Sigma}_k \mathbf{V}_i^{1/2} \, \delta_{i,j} \, \delta_{k,l}$$

hold for all $i, j, k, l \in \{0, 1, \ldots, n\}$.

If the diagonal elements of the matrix $\mathbf{V}_i$ are interpreted as *volume measures* of the sub-portfolios in accident year i, then the coordinates of $\boldsymbol{\zeta}_k$ are the expected incremental loss ratios of the sub-portfolios in development year k, which are independent of the accident years.

The multivariate additive method consists of the calculation of the estimators

$$\boldsymbol{\zeta}_k^{\mathrm{AD}} := \left(\sum_{j=0}^{n-k} \mathbf{V}_j^{1/2} \boldsymbol{\Sigma}_k^{-1} \mathbf{V}_j^{1/2} \right)^{-1} \sum_{j=0}^{n-k} \left(\mathbf{V}_j^{1/2} \boldsymbol{\Sigma}_k^{-1} \mathbf{V}_j^{1/2} \right) \mathbf{V}_j^{-1} \mathbf{Z}_{j,k}$$

of the parameters $\boldsymbol{\zeta}_k$ and of the use of the *additive predictors*

$$\mathbf{Z}_{i,k}^{\mathrm{AD}} := \mathbf{V}_i \, \boldsymbol{\zeta}_k^{\mathrm{AD}}$$

of the future incremental losses $\mathbf{Z}_{i,k}$ with $i + k \geq n + 1$. The *additive predictors* of the future cumulative losses and of the reserves are then given as usual by

$$\mathbf{S}_{i,k}^{AD} := \mathbf{S}_{i,n-i} + \sum_{l=n-i+1}^{k} \mathbf{Z}_{i,l}^{AD}$$

and

$$\mathbf{R}_i^{AD} := \sum_{l=n-i+1}^{n} \mathbf{Z}_{i,l}^{AD}$$

$$\mathbf{R}_{(c)}^{AD} := \sum_{l=c-n}^{n} \mathbf{Z}_{c-l,l}^{AD}$$

$$\mathbf{R}^{AD} := \sum_{l=1}^{n} \sum_{j=n-l+1}^{n} \mathbf{Z}_{j,l}^{AD}$$

The definitions of these predictors are by no means arbitrary: The multivariate additive model is a *linear model* and it turns out that the additive predictors are exactly the *Gauss–Markov predictors* in the multivariate additive model. Therefore all additive predictors are unbiased.

The future cumulative losses and all reserves are linear combinations of future incremental losses. If $\mathbf{Y}$ denotes any of these random vectors and $\mathbf{Y}^{AD}$ its additive predictor, then the expected squared prediction error is given by

$$E\left[\left(\mathbf{Y}^{AD} - \mathbf{Y}\right)'\left(\mathbf{Y}^{AD} - \mathbf{Y}\right)\right] = \operatorname{trace}\left(\operatorname{var}\left[\mathbf{Y}^{AD} - \mathbf{Y}\right]\right)$$

Therefore it is sufficient to determine the covariances

$$\operatorname{cov}\left[\mathbf{Z}_{i,k}^{AD} - \mathbf{Z}_{i,k}, \mathbf{Z}_{j,l}^{AD} - \mathbf{Z}_{j,l}\right]$$

for all $i, j, k, l \in \{0, 1, \ldots, n\}$ with $i + k \geq n + 1$ and $j + l \geq n + 1$, which in turn result from

$$\operatorname{cov}\left[\zeta_k^{AD}, \zeta_l^{AD}\right] = \left(\sum_{j=0}^{n-k} \mathbf{V}_j^{1/2} \Sigma_k^{-1} \mathbf{V}_j^{1/2}\right)^{-1} \delta_{k,l}$$

To estimate the matrix Σ_k with $k \in \{0, 1, \ldots, n-1\}$ one may use the estimator

$$\widetilde{\Sigma}_k := \frac{1}{n-k} \sum_{j=0}^{n-k} \mathbf{V}_j^{-1/2} \left(\mathbf{Z}_{j,k} - \mathbf{V}_j \widetilde{\zeta}_k\right)\left(\mathbf{Z}_{j,k} - \mathbf{V}_j \widetilde{\zeta}_k\right)' \mathbf{V}_j^{-1/2}$$

with

$$\widetilde{\zeta}_k := \left(\sum_{j=0}^{n-k} \mathbf{V}_j\right)^{-1} \sum_{j=0}^{n-k} \mathbf{Z}_{j,k}$$

Then $\widetilde{\zeta}_k$ is an unbiased estimator of ζ_k and its coordinates are precisely the additive estimators of the coordinates of ζ_k resulting from the application of the univariate additive method to the individual sub-portfolios. The estimator $\widetilde{\zeta}_k$ replaces the estimator ζ_k^{AD} that we actually want to use, but we are unable to do so since it depends on the matrix Σ_k to be estimated.

The estimator $\widetilde{\Sigma}_k$ is positive semidefinite with $E[\mathbf{e}_i'\widetilde{\Sigma}_k\mathbf{e}_j] \leq \mathbf{e}_i'\Sigma_k\mathbf{e}_j$ and its diagonal elements are unbiased estimators of the diagonal elements of Σ_k. The missing unbiasedness outside the diagonal is not important since the estimator has to be inverted anyway. However, invertibility presents a problem: First, since $\widetilde{\Sigma}_k$ is the sum of $n - k + 1$ matrices with rank 0 or 1, this matrix is not invertible for $k \geq n + 2 - M$; second, for $k \leq n + 1 - M$ and depending on the realizations of the incremental losses, certain terms in the sum may vanish so that for such development years the estimator $\widetilde{\Sigma}_k$ may also fail to be invertible. In the case of missing invertibility the estimator has to be modified in an appropriate manner. Finally, an estimator of Σ_n has to be determined by extrapolation in any case.

Let us finally note that the results on the multivariate additive model and the multivariate additive method remain valid when the matrices $\mathbf{V}_0, \mathbf{V}_1, \ldots, \mathbf{V}_n$ are replaced by arbitrary positive definite matrices. However, in this case the interpretation of the elements of these matrices gets lost and it is arguable whether this generalization is of practical interest.

Multivariate Panning Method

We denote by

$$\mathcal{G}_0$$

the σ-algebra generated by the family $\{\mathbf{Z}_{i,0}\}_{i\in\{0,1,\ldots,n\}}$. This σ-algebra contains the information which is available at the end of development year 0. For $i \in \{0, 1, \ldots, n\}$, we denote by

$$\mathbf{Q}_{i,0} := \mathrm{diag}(\mathbf{Z}_{i,0})$$

the diagonal matrix obtained from the random vector $\mathbf{Z}_{i,0}$.

The multivariate Panning method is based on the *multivariate Panning model*:

Multivariate Panning Model: *There exist unknown parameters $\xi_0, \xi_1, \ldots, \xi_n$ and positive definite symmetric matrices $\Sigma_0, \Sigma_1, \ldots, \Sigma_n$ such that the identities*

$$E(\mathbf{Z}_{i,k}|\mathcal{G}_0) = \mathbf{Q}_{i,0}\,\boldsymbol{\xi}_k$$
$$\mathrm{cov}(\mathbf{Z}_{i,k}, \mathbf{Z}_{j,l}|\mathcal{G}_0) = \Sigma_k\,\delta_{i,j}\,\delta_{k,l}$$

hold for all $i, j, k, l \in \{0, 1, \dots, n\}$.

The multivariate Panning method consists of the calculation of the estimators

$$\boldsymbol{\xi}_k^{\mathrm{PA}} := \left(\sum_{j=0}^{n-k} \mathbf{Q}_{j,0}\Sigma_k^{-1}\mathbf{Q}_{j,0}\right)^{-1} \sum_{j=0}^{n-k} \mathbf{Q}_{j,0}\Sigma_k^{-1}\mathbf{Z}_{j,k}$$

of the parameters $\boldsymbol{\xi}_k$ and of the use of the *Panning predictors*

$$\mathbf{Z}_{i,k}^{\mathrm{PA}} := \mathbf{Q}_{i,0}\,\boldsymbol{\xi}_k^{\mathrm{PA}}$$

of the future incremental losses $\mathbf{Z}_{i,k}$ with $i + k \geq n + 1$. Note that $\boldsymbol{\xi}_0^{\mathrm{PA}} = \mathbf{1}$. The *Panning predictors* of the future cumulative losses and of the reserves are then given as usual by the sums of the Panning predictors of the corresponding incremental losses.

The multivariate Panning model is a conditional *linear model* (under the condition $\mathcal{G}_0$) and it turns out that the Panning predictors are precisely the conditional *Gauss–Markov predictors* in the multivariate Panning model. In particular, all Panning predictors are unbiased.

Similar to the multivariate additive model, the conditional squared prediction errors of the future incremental and cumulative losses and of all reserves can be obtained from the conditional covariances

$$\mathrm{cov}\left(\mathbf{Z}_{i,k}^{\mathrm{PA}} - \mathbf{Z}_{i,k}, \mathbf{Z}_{j,l}^{\mathrm{PA}} - \mathbf{Z}_{j,l}\ \middle|\ \mathcal{G}_0\right)$$

which in turn result from

$$\mathrm{cov}\left(\boldsymbol{\xi}_k^{\mathrm{PA}}, \boldsymbol{\xi}_l^{\mathrm{PA}}\ \middle|\ \mathcal{G}_0\right) = \left(\sum_{j=0}^{n-k} \mathbf{Q}_{j,0}\Sigma_k^{-1}\mathbf{Q}_{j,0}\right)^{-1} \delta_{k,l}$$

To estimate the matrix Σ_k with $k \in \{0, 1, \dots, n - 1\}$ one may use the estimator

$$\widetilde{\Sigma}_k := \frac{1}{n - k} \sum_{j=0}^{n-k} \left(\mathbf{Z}_{j,k} - \mathbf{Q}_{j,0}\,\widetilde{\boldsymbol{\xi}}_k\right)\left(\mathbf{Z}_{j,k} - \mathbf{Q}_{j,0}\,\widetilde{\boldsymbol{\xi}}_k\right)$$

with

$$\widetilde{\boldsymbol{\xi}}_k := \left(\sum_{j=0}^{n-k} \mathbf{Q}_{j,0}^2\right)^{-1} \sum_{j=0}^{n-k} \mathbf{Q}_{j,0}\mathbf{Z}_{j,k}$$

The properties of the estimators $\widetilde{\boldsymbol{\xi}}_k$ and $\widetilde{\boldsymbol{\Sigma}}_k$ are analogous to those of the corresponding estimators in the multivariate additive method.

Multivariate Chain Ladder Method

In contrast to the multivariate additive method and the multivariate Panning method, the multivariate chain ladder method is based on cumulative losses instead of incremental ones.

For $k \in \{0, 1, \ldots, n\}$, we denote by

$$\mathcal{G}_k$$

the σ-algebra generated by the family $\{\mathbf{S}_{j,l}\}_{l \in \{0,1,\ldots,k\}, j \in \{0,1,\ldots,n-k\}}$. This σ-algebra contains the information from accident years $j \in \{0, 1, \ldots, n-k\}$ that is available at the end of development year k. For $i, k \in \{0, 1, \ldots, n\}$, we denote by

$$\mathbf{Q}_{i,k} := \mathrm{diag}(\mathbf{S}_{i,k})$$

the diagonal matrix obtained from the random vector $\mathbf{S}_{i,k}$.

The multivariate chain ladder method is based on the *multivariate chain ladder model*:

Multivariate Chain Ladder Model: *There exist unknown parameters $\varphi_1, \ldots, \varphi_n$ and positive definite symmetric matrices $\Sigma_1, \ldots, \Sigma_n$ such that the identities*

$$E(\mathbf{S}_{i,k}|\mathcal{G}_{k-1}) = \mathbf{Q}_{i,k-1}\, \varphi_k$$

$$\mathrm{cov}(\mathbf{S}_{i,k}, \mathbf{S}_{j,l}|\mathcal{G}_{k-1}) = \mathbf{Q}_{i,k-1}^{1/2}\, \Sigma_k\, \mathbf{Q}_{i,k-1}^{1/2}\, \delta_{i,j}\, \delta_{k,l}$$

hold for all $k \in \{1, \ldots, n\}$ and $i, j \in \{0, 1, \ldots, n\}$.

Passing from vectors to diagonal matrices, the first identity of the model can also be written as

$$E(\mathbf{Q}_{i,k}|\mathcal{G}_{k-1}) = \mathbf{Q}_{i,k-1}\, \Phi_k$$

where

$$\Phi_k := \mathrm{diag}(\varphi_k)$$

The multivariate chain ladder method consists in the calculation of the estimators

$$
\Phi_k^{\mathrm{CL}} := \left(\sum_{j=0}^{n-k} \mathbf{Q}_{j,k-1}^{1/2} \Sigma_k^{-1} \mathbf{Q}_{j,k-1}^{1/2} \right)^{-1} \sum_{j=0}^{n-k} \left(\mathbf{Q}_{j,k-1}^{1/2} \Sigma_k^{-1} \mathbf{Q}_{j,k-1}^{1/2} \right) \mathbf{Q}_{j,k-1}^{-1} \mathbf{Q}_{i,k}
$$

of the matrices Φ_k and of the *chain ladder predictors*

$$
\mathbf{Q}_{i,k}^{\mathrm{CL}} := \mathbf{Q}_{i,n-i} \prod_{l=n-i+1}^{k} \Phi_l^{\mathrm{CL}}
$$

of the non-observable diagonal matrices $\mathbf{Q}_{i,k}$ with $i + k \geq n + 1$.
   Letting

$$
\varphi_k^{\mathrm{CL}} := \Phi_k^{\mathrm{CL}} \mathbf{1}
$$

and

$$
\mathbf{S}_{i,k}^{\mathrm{CL}} := \mathbf{Q}_{i,k}^{\mathrm{CL}} \mathbf{1}
$$

then yields estimators of the parameters φ_k and the *chain ladder predictors* of the future cumulative losses $\mathbf{S}_{i,k}$ with $i + k \geq n + 1$. Using the chain ladder predictors of the future cumulative losses, we define the *chain ladder predictors*

$$
\mathbf{Z}_{i,k}^{\mathrm{CL}} := \begin{cases} \mathbf{S}_{i,n-i+1}^{\mathrm{CL}} - \mathbf{S}_{i,n-i} & \text{if } i + k = n + 1 \\ \mathbf{S}_{i,k}^{\mathrm{CL}} - \mathbf{S}_{i,k-1}^{\mathrm{CL}} & \text{else} \end{cases}
$$

of the future incremental losses and the chain ladder predictors of the reserves are then obtained by summation.
   The multivariate chain ladder model contains, for every development year $k \in \{1, \ldots, n\}$, a conditional *linear model* (under the condition $\mathcal{G}_{k-1}$) and it turns out that in the conditional linear model for development year k the chain ladder predictor $\mathbf{S}_{n-k+1,k}^{\mathrm{CL}}$ is precisely the conditional *Gauss–Markov predictor* of $\mathbf{S}_{n-k+1,k}$.
   To estimate the matrix Σ_k with $k \in \{0, 1, \ldots, n-1\}$ one may use the estimator

$$
\widetilde{\Sigma}_k := \frac{1}{n-k} \sum_{j=0}^{n-k} \mathbf{Q}_{j,k-1}^{-1/2} \left(\mathbf{S}_{j,k} - \mathbf{Q}_{j,k-1} \widetilde{\varphi}_k \right) \left(\mathbf{S}_{j,k} - \mathbf{Q}_{j,k-1} \widetilde{\varphi}_k \right)' \mathbf{Q}_{j,k-1}^{-1/2}
$$

with

$$
\widetilde{\varphi}_k := \left(\sum_{j=0}^{n-k} \mathbf{Q}_{j,k-1} \right)^{-1} \sum_{j=0}^{n-k} \mathbf{S}_{j,k}
$$

The properties of the estimators $\widetilde{\varphi}_k$ and $\widetilde{\Sigma}_k$ are analogous to those of the corresponding estimators in the multivariate additive method.

The calculation and the estimation of the conditional expected squared prediction error of the chain ladder predictors of reserves is afflicted with the same problems as in the univariate case.

Remarks

When using a multivariate method, prediction of the future incremental losses of a sub-portfolio is not only based on the observable incremental losses of the respective sub-portfolio, but also on the observable incremental losses of all other sub-portfolios of the total portfolio. Thus, multivariate methods use a larger data base than univariate methods do, and they also take into account the correlations between the sub-portfolios.

This advantage becomes particularly obvious if we consider prediction not only for the individual sub-portfolios but also for the total portfolio: First, prediction for the total portfolio by means of a univariate method, applied to the observable incremental or cumulative losses of the total portfolio which result from aggregation over the sub-portfolios, would ignore all correlations between the sub-portfolios as well as possibly different development speeds in the sub-portfolios. In addition, the predictors for the total portfolio and its sub-portfolios obtained by a univariate method are not consistent in general.

With regard to consistency of prediction for the total portfolio and its sub-portfolios, the Gauss–Markov property of the predictors used in the multivariate methods considered here is of particular interest: For example, under the assumptions of the multivariate additive model, the additive predictor $\mathbf{Z}_{i,k}^{\mathrm{AD}}$ of the future incremental loss $\mathbf{Z}_{i,k}$ is a Gauss–Markov predictor, and since Gauss–Markov prediction is linear, it follows that the predictor $\mathbf{1}'\mathbf{Z}_{i,k}^{\mathrm{AD}}$ of the future incremental loss $\mathbf{1}'\mathbf{Z}_{i,k}$ of the total portfolio as well as each of the predictors $\mathbf{e}_m'\mathbf{Z}_{i,k}^{\mathrm{AD}}$ of the future incremental losses $\mathbf{e}_m'\mathbf{Z}_{i,k}$ of the sub-portfolios $m \in \{1, \ldots, M\}$ is also a Gauss–Markov predictor. Thus, each of these predictors is optimal, and the identity

$$\sum_{m=1}^{M} \mathbf{e}_m' \mathbf{Z}_{i,k}^{\mathrm{AD}} = \mathbf{1}' \mathbf{Z}_{i,k}^{\mathrm{AD}}$$

shows that the predictors are consistent. The same applies, *mutatis mutandis*, to the multivariate Panning method under the assumptions of the multivariate Panning model and to the multivariate chain ladder method under the assumptions of the multivariate chain ladder model.

In practice, the applicability of multivariate methods reaches its limits when several estimators of the matrices Σ_k prove to be not invertible and hence have to be modified ad hoc, so that the quality of the method is diluted.

Notes

Keywords: Additive Method, Aggregation, Chain Ladder Method (Basics), Chain Ladder Method (Models), Chain Ladder Method (Prediction Error), Linear Models (Basics), Linear Models (Loss Reserving), Munich Chain Ladder Method, Paid & Incurred Problem, Panning Method.

References: GDV [2011], Braun [2004], Hess, Schmidt & Zocher [2006], Kremer [2005], Ludwig, Schmeißer & Thänert [2009], Ludwig & Schmidt [2010a], Pröhl & Schmidt [2005], Quarg & Mack [2004], Schmidt [2006b, 2012], Wüthrich & Merz [2008].

Munich Chain Ladder Method

Klaus D. Schmidt

The Munich Chain Ladder method is a bivariate extension of the chain ladder method. It provides a simultaneous prediction of the future cumulative losses of two data types of a portfolio taking into account the correlations between these data types.

The original formulation of the Munich Chain Ladder method refers to *paid losses* and *incurred losses* of a given portfolio, but the method as such does not take into account the particularities of these data types and is therefore not restricted to this particular application. For example, the Munich Chain Ladder method could also be applied to bodily injury claims and pure property damage claims in *motor third party liability insurance* and, in principle, it is applicable to any two sub-portfolios of a given portfolio.

We consider a portfolio consisting of two sub-portfolios and the corresponding run-off square of bivariate cumulative losses:

Accident year	Development year						
	0	1	... k	... $n-i$	... $n-1$	n	
0	$S_{0,0}$	$S_{0,1}$	... $S_{0,k}$	... $S_{0,n-i}$	... $S_{0,n-1}$	$S_{0,n}$	
1	$S_{1,0}$	$S_{1,1}$	... $S_{1,k}$	... $S_{1,n-i}$	... $S_{1,n-1}$	$S_{1,n}$	
$\vdots$	$\vdots$	$\vdots$		$\vdots$		$\vdots$	$\vdots$
i	$S_{i,0}$	$S_{i,1}$	... $S_{i,k}$	... $S_{i,n-i}$	... $S_{i,n-1}$	$S_{i,n}$	
$\vdots$	$\vdots$	$\vdots$	$\vdots$	$\vdots$		$\vdots$	$\vdots$
$n-k$	$S_{n-k,0}$	$S_{n-k,1}$	... $S_{n-k,k}$	... $S_{n-k,n-i}$	... $S_{n-k,n-1}$	$S_{n-k,n}$	
$\vdots$	$\vdots$	$\vdots$	$\vdots$	$\vdots$		$\vdots$	$\vdots$
$n-1$	$S_{n-1,0}$	$S_{n-1,1}$	... $S_{n-1,k}$	... $S_{n-1,n-i}$	... $S_{n-1,n-1}$	$S_{n-1,n}$	
n	$S_{n,0}$	$S_{n,1}$	... $S_{n,k}$	... $S_{n,n-i}$	... $S_{n,n-1}$	$S_{n,n}$	

K.D. Schmidt (✉)
Technische Universität Dresden, Dresden, Germany
e-mail: klaus.d.schmidt@tu-dresden.de

© Springer International Publishing Switzerland 2016
M. Radtke et al. (eds.), *Handbook on Loss Reserving*,
EAA Series, DOI 10.1007/978-3-319-30056-6_27

The bivariate cumulative losses have the form

$$\mathbf{S}_{i,k} = \begin{pmatrix} S_{i,k}^{(1)} \\ S_{i,k}^{(2)} \end{pmatrix}$$

where $S_{i,k}^{(1)}$ denotes the cumulative loss of data type (1) and $S_{i,k}^{(2)}$ denotes the cumulative loss of data type (2) for accident year i and development year k. We assume that the cumulative losses $\mathbf{S}_{i,k}$ are observable for $i + k \leq n$ and that they are non-observable for $i + k \geq n + 1$.

Munich Chain Ladder Model

The *Munich Chain Ladder method* is based on the *Munich Chain Ladder model*, which is a bivariate extension of the *chain ladder model of Mack* and takes into account correlations between two data types.

For $i, k \in \{0, 1, \ldots, n\}$, we denote by

$$\mathcal{F}_{i,k}$$

the σ-algebra generated by the family $\{\mathbf{S}_{i,l}\}_{l \in \{0,1,\ldots,k\}}$ and, for $r \in \{1, 2\}$ and $i, k \in \{0, 1, \ldots, n\}$, we denote by

$$\mathcal{F}_{i,k}^{(r)}$$

the σ-algebra generated by the family $\{S_{i,l}^{(r)}\}_{l \in \{0,1,\ldots,k\}}$.

The Munich Chain Ladder model consists of assumptions on the *individual development factors*

$$\widehat{\varphi}_{i,k}^{(1)} := S_{i,k}^{(1)} / S_{i,k-1}^{(1)}$$
$$\widehat{\varphi}_{i,k}^{(2)} := S_{i,k}^{(2)} / S_{i,k-1}^{(2)}$$

with $i \in \{0, 1, \ldots, n\}$ and $k \in \{1, \ldots, n\}$ and on the *individual quotients*

$$\widehat{\psi}_{i,k}^{(12)} := S_{i,k}^{(1)} / S_{i,k}^{(2)}$$
$$\widehat{\psi}_{i,k}^{(21)} := S_{i,k}^{(2)} / S_{i,k}^{(1)}$$

with $i, k \in \{0, 1, \ldots, n\}$:

Munich Chain Ladder Model:

(i) *The accident years are independent.*

(ii) *There exist unknown parameters $\lambda^{(1)}$ and $\lambda^{(2)}$ such that the identities*

$$E\left(\frac{\widehat{\varphi}_{i,k}^{(1)} - E(\widehat{\varphi}_{i,k}^{(1)}|\mathcal{F}_{i,k-1}^{(1)})}{\sqrt{\mathrm{var}(\widehat{\varphi}_{i,k}^{(1)}|\mathcal{F}_{i,k-1}^{(1)})}}\,\middle|\,\mathcal{F}_{i,k-1}\right) = \lambda^{(1)}\,\frac{\widehat{\psi}_{i,k-1}^{(21)} - E(\widehat{\psi}_{i,k-1}^{(21)}|\mathcal{F}_{i,k-1}^{(1)})}{\sqrt{\mathrm{var}(\widehat{\psi}_{i,k-1}^{(21)}|\mathcal{F}_{i,k-1}^{(1)})}}$$

$$E\left(\frac{\widehat{\varphi}_{i,k}^{(2)} - E(\widehat{\varphi}_{i,k}^{(2)}|\mathcal{F}_{i,k-1}^{(2)})}{\sqrt{\mathrm{var}(\widehat{\varphi}_{i,k}^{(2)}|\mathcal{F}_{i,k-1}^{(2)})}}\,\middle|\,\mathcal{F}_{i,k-1}\right) = \lambda^{(2)}\,\frac{\widehat{\psi}_{i,k-1}^{(12)} - E(\widehat{\psi}_{i,k-1}^{(12)}|\mathcal{F}_{i,k-1}^{(2)})}{\sqrt{\mathrm{var}(\widehat{\psi}_{i,k-1}^{(12)}|\mathcal{F}_{i,k-1}^{(2)})}}$$

hold for every $k \in \{1, \ldots, n\}$ and for all $i \in \{0, 1, \ldots, n\}$.

(iii) *There exist unknown parameters $\varphi_1^{(1)}, \ldots, \varphi_n^{(1)}$ and $\varphi_1^{(2)}, \ldots, \varphi_n^{(2)}$ as well as $\sigma_1^{(1)}, \ldots,$ $\sigma_n^{(1)}$ and $\sigma_1^{(2)}, \ldots, \sigma_n^{(2)}$ such that the identities*

$$E(\widehat{\varphi}_{i,k}^{(1)}|\mathcal{F}_{i,k-1}^{(1)}) = \varphi_k^{(1)}$$

$$E(\widehat{\varphi}_{i,k}^{(2)}|\mathcal{F}_{i,k-1}^{(2)}) = \varphi_k^{(2)}$$

and

$$\mathrm{var}(\widehat{\varphi}_{i,k}^{(1)}|\mathcal{F}_{i,k-1}^{(1)}) = (\sigma_k^{(1)})^2/S_{i,k-1}^{(1)}$$

$$\mathrm{var}(\widehat{\varphi}_{i,k}^{(2)}|\mathcal{F}_{i,k-1}^{(2)}) = (\sigma_k^{(2)})^2/S_{i,k-1}^{(2)}$$

hold for each $k \in \{1, \ldots, n\}$ and for all $i \in \{0, 1, \ldots, n\}$.

(iv) *There exist unknown parameters $\psi_0^{(12)}, \psi_1^{(12)}, \ldots, \psi_n^{(12)}$ and $\psi_0^{(21)}, \psi_1^{(21)}, \ldots,$ $\psi_n^{(21)}$ as well as $\sigma_0^{(12)}, \sigma_1^{(12)}, \ldots, \sigma_n^{(12)}$ and $\sigma_0^{(21)}, \sigma_1^{(21)}, \ldots, \sigma_n^{(21)}$ such that the identities*

$$E(\widehat{\psi}_{i,k}^{(12)}|\mathcal{F}_{i,k}^{(2)}) = \psi_k^{(12)}$$

$$E(\widehat{\psi}_{i,k}^{(21)}|\mathcal{F}_{i,k}^{(1)}) = \psi_k^{(21)}$$

and

$$\mathrm{var}(\widehat{\psi}_{i,k}^{(12)}|\mathcal{F}_{i,k}^{(2)}) = (\sigma_k^{(12)})^2/S_{i,k}^{(2)}$$

$$\mathrm{var}(\widehat{\psi}_{i,k}^{(21)}|\mathcal{F}_{i,k}^{(1)}) = (\sigma_k^{(21)})^2/S_{i,k}^{(1)}$$

hold for each $k \in \{0, 1, \ldots, n\}$ and for all $i \in \{0, 1, \ldots, n\}$.

Assumption (ii) of the Munich Chain Ladder model implies that the identities

$$\varrho(\widehat{\varphi}_{i,k}^{(1)}, \widehat{\psi}_{i,k-1}^{(21)}|\mathcal{F}_{i,k-1}^{(1)}) = \lambda^{(1)}$$

$$\varrho(\widehat{\varphi}_{i,k}^{(2)}, \widehat{\psi}_{i,k-1}^{(12)}|\mathcal{F}_{i,k-1}^{(2)}) = \lambda^{(2)}$$

hold for every $k \in \{1, \ldots, n\}$ and for all $i \in \{0, 1, \ldots, n\}$, where ϱ denotes the conditional coefficient of correlation. Moreover, it follows from assumptions (i) and (iii) that for each of the two data types the assumptions of the *chain ladder model of Mack* are fulfilled, which implies that, for every $r \in \{1, 2\}$, the parameters

$$\varphi_1^{(r)}, \ldots, \varphi_n^{(r)}$$

form development patterns for factors.

Assumption (iv) is not part of the original formulation of the Munich Chain Ladder model by Quarg and Mack (2004), but it supports the justification of the Munich Chain Ladder method by the Munich Chain Ladder model.[1]

In the following we assume that the assumptions of the Munich Chain Ladder model are fulfilled.

Using assumptions (iii) and (iv), the identities of (ii) can be written as

$$E\left(\frac{\widehat{\varphi}_{i,k}^{(1)} - \varphi_k^{(1)}}{\sigma_k^{(1)}} \,\middle|\, \mathcal{F}_{i,k-1} \right) = \lambda^{(1)} \frac{\widehat{\psi}_{i,k-1}^{(21)} - \psi_{k-1}^{(21)}}{\sigma_{k-1}^{(21)}}$$

$$E\left(\frac{\widehat{\varphi}_{i,k}^{(2)} - \varphi_k^{(2)}}{\sigma_k^{(2)}} \,\middle|\, \mathcal{F}_{i,k-1} \right) = \lambda^{(2)} \frac{\widehat{\psi}_{i,k-1}^{(12)} - \psi_{k-1}^{(12)}}{\sigma_{k-1}^{(12)}}$$

and hence in the form

$$E(\widehat{\varphi}_{i,k}^{(1)} | \mathcal{F}_{i,k-1}) = \varphi_k^{(1)} + \lambda^{(1)} \frac{\sigma_k^{(1)}}{\sigma_{k-1}^{(21)}} \left(\widehat{\psi}_{i,k-1}^{(21)} - \psi_{k-1}^{(21)} \right)$$

$$E(\widehat{\varphi}_{i,k}^{(2)} | \mathcal{F}_{i,k-1}) = \varphi_k^{(2)} + \lambda^{(2)} \frac{\sigma_k^{(2)}}{\sigma_{k-1}^{(12)}} \left(\widehat{\psi}_{i,k-1}^{(12)} - \psi_{k-1}^{(12)} \right)$$

Using the definitions of the individual development factors and those of the individual quotients yields

$$E(S_{i,k}^{(1)} | \mathcal{F}_{i,k-1}) = S_{i,k-1}^{(1)} \left(\varphi_k^{(1)} + \lambda^{(1)} \frac{\sigma_k^{(1)}}{\sigma_{k-1}^{(21)}} \left(\frac{S_{i,k-1}^{(2)}}{S_{i,k-1}^{(1)}} - \psi_{k-1}^{(21)} \right) \right)$$

$$E(S_{i,k}^{(2)} | \mathcal{F}_{i,k-1}) = S_{i,k-1}^{(2)} \left(\varphi_k^{(2)} + \lambda^{(2)} \frac{\sigma_k^{(2)}}{\sigma_{k-1}^{(12)}} \left(\frac{S_{i,k-1}^{(1)}}{S_{i,k-1}^{(2)}} - \psi_{k-1}^{(12)} \right) \right)$$

The last identities together with appropriate estimators of the parameters provide a justification of the Munich Chain Ladder method.

---

[1] Besides the parameters $\psi_k^{(21)}$ and $\psi_k^{(12)}$ as well as $\sigma_k^{(21)}$ and $\sigma_k^{(12)}$, the other parameters of the Munich Chain Ladder model (and hence those of the chain ladder model of Mack) may also be assumed to be random, but then the interpretation of the parameters

$$\varphi_1^{(r)}, \ldots, \varphi_n^{(r)}$$

as a development pattern for factors as well as the unbiasedness of the chain ladder factors gets lost.

Estimation of the Parameters

To estimate the parameters $\varphi_k^{(1)}$ and $\varphi_k^{(2)}$ with $k \in \{1, \ldots, n\}$, the Munich Chain Ladder method uses the *chain ladder factors*

$$\varphi_k^{(1)\text{CL}} := \frac{\sum_{j=0}^{n-k} S_{j,k}^{(1)}}{\sum_{j=0}^{n-k} S_{j,k-1}^{(1)}}$$

$$\varphi_k^{(2)\text{CL}} := \frac{\sum_{j=0}^{n-k} S_{j,k}^{(2)}}{\sum_{j=0}^{n-k} S_{j,k-1}^{(2)}}$$

and, to estimate the parameters $\sigma_k^{(1)}$ and $\sigma_k^{(2)}$ with $k \in \{1, \ldots, n-1\}$, it uses, like the chain ladder method, the estimators

$$\widehat{\sigma}_k^{(1)} := \left(\frac{1}{n-k} \sum_{j=0}^{n-k} S_{j,k-1}^{(1)} \left(\widehat{\varphi}_{j,k}^{(1)} - \varphi_k^{(1)\text{CL}} \right)^2 \right)^{1/2}$$

$$\widehat{\sigma}_k^{(2)} := \left(\frac{1}{n-k} \sum_{j=0}^{n-k} S_{j,k-1}^{(2)} \left(\widehat{\varphi}_{j,k}^{(2)} - \varphi_k^{(2)\text{CL}} \right)^2 \right)^{1/2}$$

Estimators $\widehat{\sigma}_n^{(1)}$ and $\widehat{\sigma}_n^{(2)}$ can then be obtained by extrapolation.

Similarly, to estimate the parameters $\psi_k^{(12)}$ and $\psi_k^{(21)}$ with $k \in \{1, \ldots, n\}$, the Munich Chain Ladder method uses the estimators

$$\widehat{\psi}_k^{(12)} := \frac{\sum_{j=0}^{n-k} S_{j,k}^{(1)}}{\sum_{j=0}^{n-k} S_{j,k}^{(2)}}$$

$$\widehat{\psi}_k^{(21)} := \frac{\sum_{j=0}^{n-k} S_{j,k}^{(2)}}{\sum_{j=0}^{n-k} S_{j,k}^{(1)}}$$

and, to estimate the parameters $\sigma_k^{(12)}$ and $\sigma_k^{(21)}$ with $k \in \{0, 1, \ldots, n-1\}$, it uses the estimators

$$\widehat{\sigma}_k^{(12)} := \left(\frac{1}{n-k} \sum_{j=0}^{n-k} S_{j,k}^{(2)} \left(\widehat{\psi}_{j,k}^{(12)} - \widehat{\psi}_k^{(12)} \right)^2 \right)^{1/2}$$

$$\widehat{\sigma}_k^{(21)} := \left(\frac{1}{n-k} \sum_{j=0}^{n-k} S_{j,k}^{(1)} \left(\widehat{\psi}_{j,k}^{(21)} - \widehat{\psi}_k^{(21)} \right)^2 \right)^{1/2}$$

Estimators of the parameters $\sigma_n^{(12)}$ and $\sigma_n^{(21)}$ are not required.

To estimate the parameters $\lambda^{(1)}$ and $\lambda^{(2)}$, the Munich Chain Ladder method uses the estimators

$$\widehat{\lambda}^{(1)} := \frac{\displaystyle\sum_{l=1}^{n}\sum_{j=0}^{n-l}\left(\frac{\widehat{\psi}_{j,l-1}^{(21)} - \widehat{\psi}_{l-1}^{(21)}}{\widehat{\sigma}_{l-1}^{(21)}}\sqrt{S_{j,l-1}^{(1)}}\right)\left(\frac{\widehat{\varphi}_{j,l}^{(1)} - \varphi_{l}^{(1)CL}}{\widehat{\sigma}_{l}^{(1)}}\sqrt{S_{j,l-1}^{(1)}}\right)}{\displaystyle\sum_{l=1}^{n}\sum_{j=0}^{n-l}\left(\frac{\widehat{\psi}_{j,l-1}^{(21)} - \widehat{\psi}_{l-1}^{(21)}}{\widehat{\sigma}_{l-1}^{(21)}}\sqrt{S_{j,l-1}^{(1)}}\right)^{2}}$$

$$\widehat{\lambda}^{(2)} := \frac{\displaystyle\sum_{l=1}^{n}\sum_{j=0}^{n-l}\left(\frac{\widehat{\psi}_{j,l-1}^{(12)} - \widehat{\psi}_{l-1}^{(12)}}{\widehat{\sigma}_{l-1}^{(12)}}\sqrt{S_{j,l-1}^{(2)}}\right)\left(\frac{\widehat{\varphi}_{j,l}^{(2)} - \varphi_{l}^{(2)CL}}{\widehat{\sigma}_{l}^{(2)}}\sqrt{S_{j,l-1}^{(2)}}\right)}{\displaystyle\sum_{l=1}^{n}\sum_{j=0}^{n-l}\left(\frac{\widehat{\psi}_{j,l-1}^{(12)} - \widehat{\psi}_{l-1}^{(12)}}{\widehat{\sigma}_{l-1}^{(12)}}\sqrt{S_{j,l-1}^{(2)}}\right)^{2}}$$

These estimators can be constructed from the identities of assumption (ii) in the following way:

- Multiply the identities by the co-factors of $\lambda^{(1)}$ respectively $\lambda^{(2)}$.
- Ignore conditioning.
- Take the sums over the run-off triangle.
- Solve the resulting equations for $\lambda^{(1)}$, respectively $\lambda^{(2)}$.
- Replace the parameters by their estimators.

Munich Chain Ladder Method

The *Munich Chain Ladder predictors* $S_{i,k}^{(1)MCL}$ and $S_{i,k}^{(2)MCL}$ of the future cumulative losses with $i, k \in \{1, \ldots, n\}$ and $i + k \geq n + 1$ are recursively defined by letting

$$S_{i,k}^{(1)MCL} := S_{i,k-1}^{(1)MCL}\left(\varphi_{k}^{(1)CL} + \widehat{\lambda}^{(1)}\frac{\widehat{\sigma}_{k}^{(1)}}{\widehat{\sigma}_{k-1}^{(21)}}\left(\frac{S_{i,k-1}^{(2)MCL}}{S_{i,k-1}^{(1)MCL}} - \widehat{\psi}_{k-1}^{(21)}\right)\right)$$

$$S_{i,k}^{(2)MCL} := S_{i,k-1}^{(2)MCL}\left(\varphi_{k}^{(2)CL} + \widehat{\lambda}^{(2)}\frac{\widehat{\sigma}_{k}^{(2)}}{\widehat{\sigma}_{k-1}^{(12)}}\left(\frac{S_{i,k-1}^{(1)MCL}}{S_{i,k-1}^{(2)MCL}} - \widehat{\psi}_{k-1}^{(12)}\right)\right)$$

with

$$S_{i,n-i}^{(1)MCL} := S_{i,n-i}^{(1)}$$
$$S_{i,n-i}^{(2)MCL} := S_{i,n-i}^{(2)}$$

Compared to the univariate chain ladder method, a correction of the chain ladder factors is carried out such that the predictors of the future cumulative losses for one type of data take into account information from the other type of data.

Remarks

The Munich Chain Ladder method was originally designed for the solution of the *paid and incurred problem* arising from the observation that in practice the predictors of the ultimate paid losses and those of the ultimate incurred losses, both determined by a univariate method, often differ from each other.

Nevertheless, like other bivariate methods which were developed to solve the paid and incurred problem, the Munich Chain Ladder method does not lead to a systematic reduction of the gap between the univariate predictors of the ultimate paid losses and those of the ultimate incurred losses. This can be seen from two examples in Quarg and Mack (2004), where such a reduction occurs for many but not all accident years.

Due to the symmetry of the Munich Chain Ladder model and due to the fact that this model does not contain any assumptions reflecting the specific relation between paid losses and incurred losses of a portfolio of risks, the Munich Chain Ladder method is applicable to any two arbitrary data types or sub-portfolios, as mentioned earlier.

Notes

Keywords: Chain Ladder Method (Basics), Chain Ladder Method (Models), Development Patterns (Basics), Multivariate Methods, Paid & Incurred Problem, Run-Off Data.

Reference: Quarg & Mack [2004].

Paid and Incurred Problem

Kathrin Bach, Klaus D. Schmidt and Katrin Thänert

The paid and incurred problem arises from the observation that in practice the predictors of the ultimate losses, determined either on the basis of *paid losses* or on that of *incurred losses*, often differ from each other. This is not too surprising since the development patterns for paid and incurred losses may be distinct and hence may result in a different development speed for each of the two types of data. Moreover, even when these development patterns are identical, the volatility of the two types of data may be distinct and thus lead to different estimators and predictors. In *long-tail* business, differences between these predictors may also result from the fact that the actual development horizon is greater than that used in the run-off squares.

For the solution of the paid and incurred problem, it is appropriate to combine the run-off data for paid and incurred losses to bivariate run-off data, to model these bivariate data taking into account the dependencies between paid and incurred losses and to predict the future paid and incurred losses simultaneously on the basis of such a bivariate model. The question is whether the gap between the predictors of paid and incurred losses can be closed or at least be reduced with such a method.

K. Bach · K.D. Schmidt (✉) · K. Thänert
Technische Universität Dresden, Dresden, Germany
e-mail: klaus.d.schmidt@tu-dresden.de

© Springer International Publishing Switzerland 2016
M. Radtke et al. (eds.), *Handbook on Loss Reserving*,
EAA Series, DOI 10.1007/978-3-319-30056-6_28

We consider the run-off square of bivariate incremental losses:

Accident year	Development year						
	0	1	... k	... $n-i$	... $n-1$	n	
0	$\mathbf{Z}_{0,0}$	$\mathbf{Z}_{0,1}$	... $\mathbf{Z}_{0,k}$	... $\mathbf{Z}_{0,n-i}$	... $\mathbf{Z}_{0,n-1}$	$\mathbf{Z}_{0,n}$	
1	$\mathbf{Z}_{1,0}$	$\mathbf{Z}_{1,1}$	... $\mathbf{Z}_{1,k}$	... $\mathbf{Z}_{1,n-i}$	... $\mathbf{Z}_{1,n-1}$	$\mathbf{Z}_{1,n}$	
$\vdots$	$\vdots$	$\vdots$	$\vdots$	$\vdots$	$\vdots$	$\vdots$	
i	$\mathbf{Z}_{i,0}$	$\mathbf{Z}_{i,1}$	... $\mathbf{Z}_{i,k}$	... $\mathbf{Z}_{i,n-i}$	... $\mathbf{Z}_{i,n-1}$	$\mathbf{Z}_{i,n}$	
$\vdots$	$\vdots$	$\vdots$	$\vdots$	$\vdots$	$\vdots$	$\vdots$	
$n-k$	$\mathbf{Z}_{n-k,0}$	$\mathbf{Z}_{n-k,1}$	... $\mathbf{Z}_{n-k,k}$	... $\mathbf{Z}_{n-k,n-i}$	... $\mathbf{Z}_{n-k,n-1}$	$\mathbf{Z}_{n-k,n}$	
$\vdots$	$\vdots$	$\vdots$	$\vdots$	$\vdots$	$\vdots$	$\vdots$	
$n-1$	$\mathbf{Z}_{n-1,0}$	$\mathbf{Z}_{n-1,1}$	... $\mathbf{Z}_{n-1,k}$	... $\mathbf{Z}_{n-1,n-i}$	... $\mathbf{Z}_{n-1,n-1}$	$\mathbf{Z}_{n-1,n}$	
n	$\mathbf{Z}_{n,0}$	$\mathbf{Z}_{n,1}$	... $\mathbf{Z}_{n,k}$	... $\mathbf{Z}_{n,n-i}$	... $\mathbf{Z}_{n,n-1}$	$\mathbf{Z}_{n,n}$	

The bivariate incremental losses have the form

$$\mathbf{Z}_{i,k} = \begin{pmatrix} Z_{i,k}^{(p)} \\ Z_{i,k}^{(q)} \end{pmatrix}$$

where $Z_{i,k}^{(p)}$ denotes the incremental paid loss and $Z_{i,k}^{(q)}$ the incremental incurred loss from accident year i in development year k. We assume that the incremental losses $\mathbf{Z}_{i,k}$ are observable for $i + k \leq n$ and that they are non-observable for $i + k \geq n + 1$.

Additive Paid and Incurred Method

We assume in this section that an additive model is given for the incremental paid losses and for the incremental incurred losses, respectively, and that the same volume measures $v_0, v_1, \ldots, v_n$ are used in both models. Then there exist unknown parameters $\zeta_0^{(p)}, \zeta_1^{(p)}, \ldots, \zeta_n^{(p)}$ and $\zeta_0^{(q)}, \zeta_1^{(q)}, \ldots, \zeta_n^{(q)}$ as well as parameters $\sigma_0^{(p)}, \sigma_1^{(p)}, \ldots, \sigma_n^{(p)}$ and $\sigma_0^{(q)}, \sigma_1^{(q)}, \ldots, \sigma_n^{(q)}$ such that the identities

$$E[Z_{i,k}^{(r)}] = v_i \, \zeta_k^{(r)}$$

$$\text{var}[Z_{i,k}^{(r)}] = v_i \, (\sigma_k^{(r)})^2$$

hold for all $r \in \{p, q\}$ and all $i, k \in \{0, 1, \ldots, n\}$. This means that, for every $r \in \{p, q\}$, there exists a linear model of the form

$$E\left[\begin{pmatrix} \mathbf{X}_1^{(r)} \\ \mathbf{X}_2^{(r)} \end{pmatrix}\right] = \begin{pmatrix} \mathbf{B}_1 \\ \mathbf{B}_2 \end{pmatrix} \boldsymbol{\zeta}^{(r)}$$

$$\mathrm{var}\left[\begin{pmatrix} \mathbf{X}_1^{(r)} \\ \mathbf{X}_2^{(r)} \end{pmatrix}\right] = \begin{pmatrix} \boldsymbol{\Sigma}_{11}^{(r)} & \mathbf{O} \\ \mathbf{O} & \boldsymbol{\Sigma}_{22}^{(r)} \end{pmatrix}$$

such that $\mathbf{X}_1^{(r)}$ contains the observable incremental losses and $\mathbf{X}_2^{(r)}$ contains the non-observable ones.

In order to combine these two linear models into a bivariate linear model, we also need assumptions about the covariances $\mathrm{cov}[\mathbf{X}_i^{(p)}, \mathbf{X}_j^{(q)}]$ with $i, j \in \{1, 2\}$ and $i \neq j$. To simplify the representation, we assume in the following that these covariances are equal to $\mathbf{O}$. We thus obtain the bivariate linear model

$$E\left[\begin{pmatrix} \mathbf{X}_1^{(p)} \\ \mathbf{X}_1^{(q)} \\ \mathbf{X}_2^{(p)} \\ \mathbf{X}_2^{(q)} \end{pmatrix}\right] = \begin{pmatrix} \mathbf{B}_1 & \mathbf{O} \\ \mathbf{O} & \mathbf{B}_1 \\ \mathbf{B}_2 & \mathbf{O} \\ \mathbf{O} & \mathbf{B}_2 \end{pmatrix} \begin{pmatrix} \boldsymbol{\zeta}^{(p)} \\ \boldsymbol{\zeta}^{(q)} \end{pmatrix}$$

$$\mathrm{var}\left[\begin{pmatrix} \mathbf{X}_1^{(p)} \\ \mathbf{X}_1^{(q)} \\ \mathbf{X}_2^{(p)} \\ \mathbf{X}_2^{(q)} \end{pmatrix}\right] = \begin{pmatrix} \boldsymbol{\Sigma}_{11}^{(p)} & \mathbf{O} & \mathbf{O} & \mathbf{O} \\ \mathbf{O} & \boldsymbol{\Sigma}_{11}^{(q)} & \mathbf{O} & \mathbf{O} \\ \mathbf{O} & \mathbf{O} & \boldsymbol{\Sigma}_{22}^{(p)} & \mathbf{O} \\ \mathbf{O} & \mathbf{O} & \mathbf{O} & \boldsymbol{\Sigma}_{22}^{(q)} \end{pmatrix}$$

With regard to the paid and incurred problem it is appropriate to impose the condition that the expected ultimate paid losses and the expected ultimate incurred losses are equal for every accident year. Since the ultimate losses satisfy

$$E[S_{i,n}^{(r)}] = v_i \sum_{l=0}^{n} \boldsymbol{\zeta}_l^{(r)}$$

the condition yields the *additive paid and incurred constraint*

$$\sum_{l=0}^{n} \boldsymbol{\zeta}_l^{(p)} = \sum_{l=0}^{n} \boldsymbol{\zeta}_l^{(q)}$$

on the parameters. This constraint can be written in the form $\mathbf{1}'\boldsymbol{\zeta}^{(p)} = \mathbf{1}'\boldsymbol{\zeta}^{(q)}$ and hence also in the form

$$\mathbf{H}\begin{pmatrix} \boldsymbol{\zeta}^{(p)} \\ \boldsymbol{\zeta}^{(q)} \end{pmatrix} = \mathbf{0}$$

with

$$\mathbf{H} := \begin{pmatrix} \mathbf{1}' & -\mathbf{1}' \end{pmatrix}$$

Letting

$$\mathbf{X}_s := \begin{pmatrix} \mathbf{X}_s^{(p)} \\ \mathbf{X}_s^{(q)} \end{pmatrix}$$

$$\mathbf{A}_s := \begin{pmatrix} \mathbf{B}_s & \mathbf{O} \\ \mathbf{O} & \mathbf{B}_s \end{pmatrix}$$

$$\Sigma_{ss} := \begin{pmatrix} \Sigma_{ss}^{(p)} & \mathbf{O} \\ \mathbf{O} & \Sigma_{ss}^{(q)} \end{pmatrix}$$

for $s \in \{1, 2\}$ as well as

$$\boldsymbol{\zeta} := \begin{pmatrix} \boldsymbol{\zeta}^{(p)} \\ \boldsymbol{\zeta}^{(q)} \end{pmatrix}$$

we thus obtain the linear model

$$E\left[\begin{pmatrix} \mathbf{X}_1 \\ \mathbf{X}_2 \end{pmatrix}\right] = \begin{pmatrix} \mathbf{A}_1 \\ \mathbf{A}_2 \end{pmatrix} \boldsymbol{\zeta}$$

$$\operatorname{var}\left[\begin{pmatrix} \mathbf{X}_1 \\ \mathbf{X}_2 \end{pmatrix}\right] = \begin{pmatrix} \Sigma_{11} & \mathbf{O} \\ \mathbf{O} & \Sigma_{22} \end{pmatrix}$$

with the constraint

$$\mathbf{H}\boldsymbol{\zeta} = \mathbf{0}$$

on the parameter. Define

$$\mathbf{G} := (\mathbf{H}(\mathbf{A}_1'\Sigma_{11}^{-1}\mathbf{A}_1)^{-1}\mathbf{H}')^{-1}$$

and let $\boldsymbol{\zeta}^{\mathrm{GM}}$ denote the Gauss–Markov estimator of $\boldsymbol{\zeta}$ in the linear model without the additive paid and incurred constraint $\mathbf{H}\boldsymbol{\zeta} = \mathbf{0}$. Then the $(\mathbf{H}, \mathbf{0})$-Gauss–Markov estimator $\boldsymbol{\zeta}_{\mathbf{H},\mathbf{0}}^{\mathrm{GM}}$ of $\boldsymbol{\zeta}$ satisfies

$$\boldsymbol{\zeta}_{\mathbf{H},\mathbf{0}}^{\mathrm{GM}} = \boldsymbol{\zeta}^{\mathrm{GM}} - (\mathbf{A}_1'\Sigma_{11}^{-1}\mathbf{A}_1)^{-1}\mathbf{H}'\mathbf{G}\mathbf{H}\boldsymbol{\zeta}^{\mathrm{GM}}$$

and the $(\mathbf{H}, \mathbf{0})$-Gauss–Markov predictor $\mathbf{X}_{2,\mathbf{H},\mathbf{0}}^{\mathrm{GM}}$ of $\mathbf{X}_2$ satisfies

$$\mathbf{X}_{2;\mathbf{H},\mathbf{0}}^{\mathrm{GM}} = \mathbf{A}_2\boldsymbol{\zeta}_{\mathbf{H},\mathbf{0}}^{\mathrm{GM}}$$

Letting

$$\mathbf{G}^{(r)} := (\mathbf{B}_1'(\Sigma_{11}^{(r)})^{-1}\mathbf{B}_1)^{-1}$$

for $r \in \{p, q\}$ we obtain

$$(\mathbf{A}_1' \Sigma_{11}^{-1} \mathbf{A}_1)^{-1} = \begin{pmatrix} \mathbf{G}^{(p)} & \mathbf{O} \\ \mathbf{O} & \mathbf{G}^{(q)} \end{pmatrix}$$

and hence

$$\mathbf{G} = (\mathbf{1}'(\mathbf{G}^{(p)} + \mathbf{G}^{(q)})\mathbf{1})^{-1}$$

as well as

$$(\mathbf{A}_1' \Sigma_{11}^{-1} \mathbf{A}_1)^{-1} \mathbf{H}' \mathbf{H} = \begin{pmatrix} \mathbf{G}^{(p)} \mathbf{1} \mathbf{1}' & -\mathbf{G}^{(q)} \mathbf{1} \mathbf{1}' \\ -\mathbf{G}^{(q)} \mathbf{1} \mathbf{1}' & \mathbf{G}^{(p)} \mathbf{1} \mathbf{1}' \end{pmatrix}$$

Therefore, the $(\mathbf{H}, \mathbf{0})$-Gauss–Markov estimator $\boldsymbol{\zeta}_{\mathbf{H},0}^{\mathrm{GM}}$ of $\boldsymbol{\zeta}$ satisfies

$$\begin{pmatrix} \boldsymbol{\zeta}_{\mathbf{H},0}^{(p)\mathrm{GM}} \\ \boldsymbol{\zeta}_{\mathbf{H},0}^{(q)\mathrm{GM}} \end{pmatrix} = \begin{pmatrix} \boldsymbol{\zeta}^{(p)\mathrm{GM}} \\ \boldsymbol{\zeta}^{(q)\mathrm{GM}} \end{pmatrix} - \frac{1}{\mathbf{1}'(\mathbf{G}^{(p)} + \mathbf{G}^{(q)})\mathbf{1}} \begin{pmatrix} \mathbf{G}^{(p)} \mathbf{1} \mathbf{1}' (\boldsymbol{\zeta}^{(p)\mathrm{GM}} - \boldsymbol{\zeta}^{(q)\mathrm{GM}}) \\ \mathbf{G}^{(q)} \mathbf{1} \mathbf{1}' (\boldsymbol{\zeta}^{(q)\mathrm{GM}} - \boldsymbol{\zeta}^{(p)\mathrm{GM}}) \end{pmatrix}$$

Because of $\mathbf{1}'(\mathbf{G}^{(p)} + \mathbf{G}^{(q)})\mathbf{1} > 0$ as well as $\mathbf{G}^{(p)} \mathbf{1} > \mathbf{0}$ and $\mathbf{G}^{(q)} \mathbf{1} > \mathbf{0}$ (by coordinates) the difference between the Gauss–Markov estimators of $\boldsymbol{\zeta}$ with or without the additive paid and incurred constraint depends only on the difference between $\mathbf{1}' \boldsymbol{\zeta}^{(p)\mathrm{GM}}$ and $\mathbf{1}' \boldsymbol{\zeta}^{(q)\mathrm{GM}}$.

- In the case $\mathbf{1}' \boldsymbol{\zeta}^{(p)\mathrm{GM}} = \mathbf{1}' \boldsymbol{\zeta}^{(q)\mathrm{GM}}$ one has

$$\boldsymbol{\zeta}_{\mathbf{H},0}^{\mathrm{GM}} = \boldsymbol{\zeta}^{\mathrm{GM}}$$

which means that in this case the additive paid and incurred method does not result in any change of the predictors of the ultimate losses.
- In the case $\mathbf{1}' \boldsymbol{\zeta}^{(p)\mathrm{GM}} < \mathbf{1}' \boldsymbol{\zeta}^{(q)\mathrm{GM}}$ one has

$$\boldsymbol{\zeta}_{\mathbf{H},0}^{(p)\mathrm{GM}} > \boldsymbol{\zeta}^{(p)\mathrm{GM}}$$
$$\boldsymbol{\zeta}_{\mathbf{H},0}^{(q)\mathrm{GM}} < \boldsymbol{\zeta}^{(q)\mathrm{GM}}$$

and hence

$$\mathbf{X}_{2;\mathbf{H},0}^{(p)\mathrm{GM}} > \mathbf{X}_2^{(p)\mathrm{GM}}$$
$$\mathbf{X}_{2;\mathbf{H},0}^{(q)\mathrm{GM}} < \mathbf{X}_2^{(q)\mathrm{GM}}$$

In this case we thus obtain

$$\mathbf{X}_{2;\mathbf{H},0}^{(q)\mathrm{GM}} - \mathbf{X}_{2;\mathbf{H},0}^{(p)\mathrm{GM}} < \mathbf{X}_2^{(q)\mathrm{GM}} - \mathbf{X}_2^{(p)\mathrm{GM}}$$

Thus, if the left-hand side of this inequality is positive, then the additive paid and incurred method reduces the gap between the predictors of the future paid and incurred incremental losses, and this also applies to the predictors of all reserves and ultimate losses.

Because of

$$\mathbf{1}'\boldsymbol{\zeta}^{(q)\mathrm{GM}} - \mathbf{1}'\boldsymbol{\zeta}^{(p)\mathrm{GM}} = \sum_{l=0}^{n}(\zeta_l^{(q)\mathrm{GM}} - \zeta_l^{(p)\mathrm{GM}}) = \sum_{l=0}^{n}\frac{\sum_{j=0}^{n-l}(Z_{j,l}^{(q)} - Z_{j,l}^{(p)})}{\sum_{j=0}^{n-l} v_j}$$

the difference between $\mathbf{1}'\boldsymbol{\zeta}^{(q)\mathrm{GM}}$ and $\mathbf{1}'\boldsymbol{\zeta}^{(p)\mathrm{GM}}$ is determined by the increments $Z_{j,l}^{(q)} - Z_{j,l}^{(p)}$ of the *individual loss reserves*.

Panning Paid and Incurred Method

We assume in this section that a Panning model is given for the incremental paid losses and for the incremental incurred losses, respectively. Then there exist unknown parameters $\xi_0^{(p)}, \xi_1^{(p)}, \ldots, \xi_n^{(p)}$ and $\xi_0^{(q)}, \xi_1^{(q)}, \ldots, \xi_n^{(q)}$ as well as parameters $\sigma_0^{(p)}, \sigma_1^{(p)}, \ldots, \sigma_n^{(p)}$ and $\sigma_0^{(q)}, \sigma_1^{(q)}, \ldots, \sigma_n^{(q)}$ such that the identities

$$E(Z_{i,k}^{(r)}|\mathcal{G}_0^{(r)}) = Z_{i,0}^{(r)} \xi_k^{(r)}$$
$$\mathrm{var}(Z_{i,k}^{(r)}|\mathcal{G}_0^{(r)}) = (\sigma_k^{(r)})^2$$

hold for all $r \in \{p, q\}$ and all $i, k \in \{0, 1, \ldots, n\}$, where

$$\mathcal{G}_0^{(r)}$$

denotes the σ-algebra generated by the family $\{Z_{i,0}^{(r)}\}_{i\in\{0,1,\ldots,n\}}$. This means that, for every $r \in \{p, q\}$, there exists a conditional linear model of the form

$$E\left(\begin{pmatrix}\mathbf{X}_1^{(r)}\\\mathbf{X}_2^{(r)}\end{pmatrix}\middle|\mathcal{G}_0^{(r)}\right) = \begin{pmatrix}\mathbf{B}_1^{(r)}\\\mathbf{B}_2^{(r)}\end{pmatrix}\boldsymbol{\xi}^{(r)}$$
$$\mathrm{var}\left(\begin{pmatrix}\mathbf{X}_1^{(r)}\\\mathbf{X}_2^{(r)}\end{pmatrix}\middle|\mathcal{G}_0^{(r)}\right) = \begin{pmatrix}\Sigma_{11}^{(r)} & \mathbf{O}\\\mathbf{O} & \Sigma_{22}^{(r)}\end{pmatrix}$$

such that $\mathbf{X}_1^{(r)}$ contains the observable incremental losses and $\mathbf{X}_2^{(r)}$ contains the non-observable ones.

In order to combine these two linear models into a bivariate linear model, we replace the σ-algebras $\mathcal{G}_0^{(p)}$ and $\mathcal{G}_0^{(q)}$ by the σ-algebra $\mathcal{G}_0$ generated by the family

$\{\mathbf{Z}_{i,0}\}_{i\in\{0,1,\dots,n\}}$ and then proceed as in the case of the additive paid and incurred method.

In analogy with the additive paid and incurred method, a natural condition to be imposed would be the requirement that the conditional expectations of the ultimate paid and incurred losses are equal for every accident year. This condition, however, is too strong: Since the ultimate losses satisfy

$$E(S_{i,n}^{(r)}|\mathcal{G}_0) = Z_{i,0}^{(r)} \sum_{l=0}^{n} \xi_l^{(r)}$$

this condition results in the constraint that the identity

$$Z_{i,0}^{(p)} \sum_{l=0}^{n} \xi_l^{(p)} = Z_{i,0}^{(q)} \sum_{l=0}^{n} \xi_l^{(q)}$$

holds for all $i \in \{0, 1, \dots, n\}$, which is impossible unless the ratios $Z_{i,0}^{(p)}/Z_{i,0}^{(q)}$ are non-random and identical for all accident years. We therefore replace this requirement by the weaker condition that only the sums over all accident years of the conditional expectations of the ultimate paid and incurred losses are equal. This leads to the *Panning paid and incurred constraint*

$$\sum_{j=0}^{n} Z_{j,0}^{(p)} \sum_{l=0}^{n} \xi_l^{(p)} = \sum_{j=0}^{n} Z_{j,0}^{(q)} \sum_{l=0}^{n} \xi_l^{(q)}$$

on the parameters. Since this constraint can be written in the form

$$\mathbf{H\xi} = \mathbf{0}$$

with

$$\mathbf{H} := \left(\sum_{j=0}^{n} Z_{j,0}^{(p)} \mathbf{1}' \quad -\sum_{j=0}^{n} Z_{j,0}^{(q)} \mathbf{1}' \right)$$

we also obtain in this case a linear model with a constraint on the parameter.

Remarks

In the additive paid and incurred model and in the Panning paid and incurred model, the assumptions on the paid losses and those on the incurred losses are formally identical, and this is also the case in the *Munich Chain Ladder model*. This symmetry is problematic, as the incurred losses contain the paid losses. Models for the paid losses and the individual loss reserves, which add up to the incurred losses, would be preferable.

Unlike the Munich Chain Ladder model, which is not only applicable to paid and incurred losses, the additive paid and incurred model as well as the Panning paid and incurred model contains an assumption which relates the two types of data and is specific to the paid and incurred problem.

Despite these specific assumptions for the paid and incurred problem, neither the additive paid and incurred method nor the Panning paid and incurred method results in a systematic reduction of the gap between the corresponding univariate predictors of the ultimate paid and incurred losses. As in the Munich Chain Ladder method, in both cases it depends on the data whether this desired effect appears or not.

The paid and incurred problem has already been discussed by Halliwell (1997).

Notes

Keywords: Additive Method, Linear Models (Basics), Linear Models (Loss Reserving), Multivariate Methods, Munich Chain Ladder Method, Panning Method, Run-Off Data, Tail Estimation.

References: Halliwell [1997], Kloberdanz & Schmidt [2008, 2009], Ludwig, Schmeißer & Thänert [2009], Quarg & Mack [2004], Schmidt [2012].

Panning Method

Alexander Ludwig and Klaus D. Schmidt

Consider the run-off square of incremental losses:

Accident year	Development year							
	0	1	... k	... $n-i$	... $n-1$	n
0	$Z_{0,0}$	$Z_{0,1}$	... $Z_{0,k}$	... $Z_{0,n-i}$	... $Z_{0,n-1}$	$Z_{0,n}$
1	$Z_{1,0}$	$Z_{1,1}$	... $Z_{1,k}$	... $Z_{1,n-i}$	... $Z_{1,n-1}$	$Z_{1,n}$
$\vdots$	$\vdots$	$\vdots$	$\vdots$	$\vdots$	$\vdots$	$\vdots$
i	$Z_{i,0}$	$Z_{i,1}$	... $Z_{i,k}$	... $Z_{i,n-i}$	... $Z_{i,n-1}$	$Z_{i,n}$
$\vdots$	$\vdots$	$\vdots$	$\vdots$	$\vdots$	$\vdots$	$\vdots$
$n-k$	$Z_{n-k,0}$	$Z_{n-k,1}$	... $Z_{n-k,k}$	... $Z_{n-k,n-i}$	... $Z_{n-k,n-1}$	$Z_{n-k,n}$
$\vdots$	$\vdots$	$\vdots$	$\vdots$	$\vdots$	$\vdots$	$\vdots$
$n-1$	$Z_{n-1,0}$	$Z_{n-1,1}$	... $Z_{n-1,k}$	... $Z_{n-1,n-i}$	... $Z_{n-1,n-1}$	$Z_{n-1,n}$
n	$Z_{n,0}$	$Z_{n,1}$	... $Z_{n,k}$	... $Z_{n,n-i}$	... $Z_{n,n-1}$	$Z_{n,n}$

We assume that the incremental losses $Z_{i,k}$ are observable for $i + k \leq n$ and that they are non-observable for $i + k \geq n + 1$. For $i, k \in \{0, 1, \ldots, n\}$ we denote by

$$S_{i,k} := \sum_{l=0}^{k} Z_{i,l}$$

the cumulative loss from accident year i in development year k.

The Panning method is a prediction method based on the development pattern of Panning:

A. Ludwig · K.D. Schmidt (✉)
Technische Universität Dresden, Dresden, Germany
e-mail: klaus.d.schmidt@tu-dresden.de

© Springer International Publishing Switzerland 2016
M. Radtke et al. (eds.), *Handbook on Loss Reserving*,
EAA Series, DOI 10.1007/978-3-319-30056-6_29

Development Pattern of Panning: *There exist parameters $\xi_0, \xi_1, \ldots, \xi_n$ with $\xi_0 = 1$ such that the identity*

$$\frac{E[Z_{i,k}]}{E[Z_{i,0}]} = \xi_k$$

holds for all $k \in \{0, 1, \ldots, n\}$ and for all $i \in \{0, 1, \ldots, n\}$.

In this article we assume that a development pattern of Panning exists. Then the parameters $\vartheta_0, \vartheta_1, \ldots, \vartheta_n$ with

$$\vartheta_k := \frac{\xi_k}{\sum_{l=0}^{n} \xi_l}$$

form a *development pattern for incremental quotas* and the parameters $\gamma_0, \gamma_1, \ldots, \gamma_n$ with

$$\gamma_k := \frac{\sum_{l=0}^{k} \xi_l}{\sum_{l=0}^{n} \xi_l}$$

form a *development pattern for quotas*. In particular, we are in the situation of a *multiplicative model*.

The Panning method consists of two steps:

- For every development year $k \in \{0, 1, \ldots, n\}$, the parameter ξ_k is estimated by the *Panning factor*

$$\xi_k^{\text{PA}} := \frac{\sum_{j=0}^{n-k} Z_{j,0} Z_{j,k}}{\sum_{j=0}^{n-k} Z_{j,0}^2}$$

(such that $\xi_0^{\text{PA}} = 1$). Because of

$$\xi_k^{\text{PA}} = \sum_{j=0}^{n-k} \frac{Z_{j,0}^2}{\sum_{h=0}^{n-k} Z_{h,0}^2} \frac{Z_{j,k}}{Z_{j,0}}$$

the Panning factor ξ_k^{PA} is a weighted mean of the observable individual incremental development factors $Z_{i,k}/Z_{i,0}$ of development year k with weights being proportional to the squared incremental losses of development year 0.

- For every accident year i and every development year k such that $i + k \geq n + 1$, the future incremental loss $Z_{i,k}$ is predicted by the *Panning predictor*

$$Z_{i,k}^{\text{PA}} := Z_{i,0} \, \xi_k^{\text{PA}}$$

The definition of the Panning predictors of the incremental losses replicates the identity

$$E[Z_{i,k}] = E[Z_{i,0}] \frac{E[Z_{i,k}]}{E[Z_{i,0}]} = E[Z_{i,0}] \xi_k$$

which results from the development pattern of Panning.

Using the Panning predictors of the future incremental losses, we define the *Pan-ning predictors*

$$S_{i,k}^{PA} := S_{i,n-i} + \sum_{l=n-i+1}^{k} Z_{i,l}^{PA} = S_{i,n-i} + Z_{i,0} \sum_{l=n-i+1}^{k} \xi_l^{PA}$$

of the future cumulative losses $S_{i,k}$ and the *Panning predictors*

$$R_i^{PA} := \sum_{l=n-i+1}^{n} Z_{i,l}^{PA}$$

$$R_{(c)}^{PA} := \sum_{l=c-n}^{n} Z_{c-l,l}^{PA}$$

$$R^{PA} := \sum_{l=1}^{n} \sum_{j=n-l+1}^{n} Z_{j,l}^{PA}$$

of the accident year reserves R_i, with $i \in \{1, \ldots, n\}$, of the calendar year reserves $R_{(c)}$, with $c \in \{n+1, \ldots, 2n\}$, and of the aggregate loss reserve R. These reserve predictors are also called *Panning reserves*.

Example A. Calculation of the Panning predictors of the incremental losses:

Accident year i	Development year k						Sum
	0	1	2	3	4	5	
0	1001	854	568	565	347	148	
1	1113	990	671	648	422	165	
2	1265	1168	800	744	461	187	
3	1490	1383	1007	864	543	220	
4	1725	1536	1089	1001	629	255	
5	1889	1702	1193	1096	689	279	
ξ_k^{PA}	1	0.90	0.63	0.58	0.36	0.15	3.63
ϑ_k^{PA}	0.28	0.25	0.17	0.16	0.10	0.04	1
γ_k^{PA}	0.28	0.52	0.70	0.86	0.96	1	

Reserves:

Accident year i	Reserve R_i^{PA}
1	165
2	648
3	1628
4	2974
5	4959
sum	10374

Calendar year c	Reserve $R_{(c)}^{PA}$
6	4282
7	2924
8	1945
9	944
10	279
sum	10374

The estimators of the development pattern for quotas and incremental quotas are not required for the Panning method and serve only for comparison with other methods.

Example B. In this example the incremental loss $Z_{4,1}$ is increased by 1000:

Accident year i	Development year k						Sum
	0	1	2	3	4	5	
0	1001	854	568	565	347	148	
1	1113	990	671	648	422	165	
2	1265	1168	800	744	461	187	
3	1490	1383	1007	864	543	220	
4	1725	2536	1089	1001	629	255	
5	1889	2063	1193	1096	689	279	
ξ_k^{PA}	1	1.09	0.63	0.58	0.36	0.15	3.82
ϑ_k^{PA}	0.26	0.29	0.17	0.15	0.10	0.04	1
γ_k^{PA}	0.26	0.55	0.71	0.87	0.96	1	

Reserves:

Accident year i	Reserve R_i^{PA}
1	165
2	648
3	1628
4	2974
5	5320
sum	10735

Calendar year c	Reserve $R_{(c)}^{PA}$
6	4643
7	2924
8	1945
9	944
10	279
sum	10735

The outlier $Z_{4,1}$ affects the estimator of the parameter ξ_1 and hence the predictors of the incremental loss $Z_{5,1}$ and of the reserves R_5 and $R_{(6)}$.

Bornhuetter–Ferguson Principle

Letting

$$\gamma_k^{\mathrm{PA}} := \frac{\sum_{l=0}^{k} \xi_l^{\mathrm{PA}}}{\sum_{l=0}^{n} \xi_l^{\mathrm{PA}}} \quad \text{and} \quad \alpha_i^{\mathrm{PA}} := Z_{i,0} \sum_{l=0}^{n} \xi_l^{\mathrm{PA}}$$

the Panning predictors of the future cumulative losses can be represented in the form

$$S_{i,k}^{\mathrm{PA}} = S_{i,n-i} + \left(\gamma_k^{\mathrm{PA}} - \gamma_{n-i}^{\mathrm{PA}} \right) \alpha_i^{\mathrm{PA}}$$

Therefore, the Panning method is subject to the *Bornhuetter–Ferguson principle*.

Linear Model

A development pattern of Panning exists whenever the assumptions of the *Panning model* are fulfilled. Let

$$\mathcal{G}_0$$

denote the σ-algebra generated by the family $\{Z_{i,0}\}_{i \in \{0,1,\dots,n\}}$ and hence by the information of development year 0.

> **Panning Model**: *There exist unknown parameters $\xi_0, \xi_1, \dots, \xi_n$ and parameters $\sigma_0^2, \sigma_1^2, \dots,$ σ_n^2 such that the identities*
>
> $$E(Z_{i,k}|\mathcal{G}_0) = Z_{i,0}\,\xi_k$$
> $$\mathrm{cov}(Z_{i,k}, Z_{j,l}|\mathcal{G}_0) = \sigma_k^2\,\delta_{i,j}\,\delta_{k,l}$$
>
> *hold for all $i, j, k, l \in \{0, 1, \dots, n\}$.*

The Panning model is a *conditional linear model* under the condition $\mathcal{G}_0$ and it is obvious that all Panning predictors are linear combinations of the observable incremental losses with $\mathcal{G}_0$-measurable coefficients. Further properties of the Panning predictors result from the theory of linear models:

> **Theorem.** *In the Panning model, the Panning predictor of the future incremental loss $Z_{i,k}$ is conditionally unbiased, it is optimal in the sense that it minimizes the conditional expected squared prediction error*
>
> $$E\left(\left(\widehat{Z}_{i,k} - Z_{i,k} \right)^2 \Big| \mathcal{G}_0 \right)$$
>
> *over all conditionally unbiased linear predictors $\widehat{Z}_{i,k}$ of $Z_{i,k}$, and it is the only predictor having this property. These properties also hold for the Panning predictors of cumulative losses and reserves.*

Under the assumptions of the Panning model, the theorem asserts that the Panning predictors are precisely the *conditional Gauss–Markov predictors*. In particular, it is possible to determine the expected squared prediction errors of the Panning reserves and one obtains

$$E\left(\left(R_i^{\mathrm{PA}} - R_i\right)^2 \big| \mathcal{G}_0\right)$$

$$= Z_{i,0}^2 \sum_{l=n-i+1}^{n} \left(\frac{1}{\sum_{h=0}^{n-l} Z_{h,0}^2} + \frac{1}{Z_{i,0}^2}\right)\sigma_l^2$$

$$E\left(\left(R_{(c)}^{\mathrm{PA}} - R_{(c)}\right)^2 \big| \mathcal{G}_0\right)$$

$$= \sum_{l=c-n}^{n} Z_{c-1,0}^2 \left(\frac{1}{\sum_{h=0}^{n-l} Z_{h,0}^2} + \frac{1}{Z_{c-1,0}^2}\right)\sigma_l^2$$

$$E\left(\left(R^{\mathrm{PA}} - R\right)^2 \big| \mathcal{G}_0\right)$$

$$= \sum_{l=1}^{n} \left(\sum_{j=n-l+1}^{n} Z_{j,0}\right)^2 \left(\frac{1}{\sum_{h=0}^{n-l} Z_{h,0}^2} + \frac{l}{(\sum_{h=n-l+1}^{n} Z_{h,0})^2}\right)\sigma_l^2$$

To estimate the prediction error it is sufficient to replace the variance parameters $\sigma_1^2, \ldots, \sigma_n^2$, which appear in this formulae, by the suitable estimators. Usually the unbiased estimators

$$\widehat{\sigma}_k^2 := \frac{1}{n-k} \sum_{j=0}^{n-k} \left(Z_{j,k} - Z_{j,0}\, \xi_k^{\mathrm{PA}}\right)^2$$

are chosen for $k \in \{1, \ldots, n-1\}$ and an estimator $\widehat{\sigma}_n^2$ is determined by extrapolation.

Remarks

The structure of the Panning method is very similar to that of the *additive method* and that of the *chain ladder method*. Correspondingly, there exists a strong similarity between the Panning model and the *additive model* and the *chain ladder model of Schnaus*.

Variations of the model can be obtained by replacing the condition

$$\mathrm{cov}(Z_{i,k}, Z_{j,l}|\mathcal{G}_0) = \sigma_k^2\, \delta_{i,j}\, \delta_{k,l}$$

with a condition of the form

$$\mathrm{cov}(Z_{i,k}, Z_{j,l}|\mathcal{G}_0) = w_i\, \sigma_k^2\, \delta_{i,j}\, \delta_{k,l}$$

with an accident year factor w_i. For example, by analogy with the additive model and with the chain ladder model of Schnaus, one may choose $w_i := Z_{i,0}$. Of course, the theory of linear models is also applicable to such variations of the Panning model, and instead of the Panning estimators

$$\xi_k^{\mathrm{PA}} = \sum_{j=0}^{n-k} \frac{Z_{j,0}^2}{\sum_{h=0}^{n-k} Z_{h,0}^2} \frac{Z_{j,k}}{Z_{j,0}}$$

one would then obtain the estimators

$$\widetilde{\xi}_k = \sum_{j=0}^{n-k} \frac{Z_{j,0}^2/w_j}{\sum_{h=0}^{n-k} Z_{h,0}^2/w_h} \frac{Z_{j,k}}{Z_{j,0}}$$

as Gauss–Markov estimators of the parameters, and accordingly one would obtain other Gauss–Markov predictors and other unbiased estimators of the variance parameters.

Finally, the Panning method can be varied by choosing different weights in the Panning estimators

$$\xi_k^{\mathrm{PA}} = \sum_{j=0}^{n-k} \frac{Z_{j,0}^2}{\sum_{h=0}^{n-k} Z_{h,0}^2} \frac{Z_{j,k}}{Z_{j,0}}$$

This, however, corresponds to the use of appropriate accident year factors in the variance condition of the Panning model.

Notes

Keywords: Additive Method, Bornhuetter–Ferguson Method, Bornhuetter–Ferguson Principle, Chain Ladder Method (Basics), Development Patterns (Basics), Development Patterns (Estimation), Linear Models (Basics), Linear Models (Loss Reserving), Multiplicative Models, Multivariate Methods, Paid & Incurred Problem, Run-Off Triangles.

References: Ludwig, Schmeißer & Thänert [2009], Ludwig & Schmidt [2010b], Panning [2006], Schmidt [2012], Schmidt & Zocher [2008].

Poisson Model

Klaus D. Schmidt

Consider the run-off square of incremental losses:

Accident year	Development year						
	0	1	... k	... $n-i$	... $n-1$	n	
0	$Z_{0,0}$	$Z_{0,1}$	... $Z_{0,k}$	$Z_{0,n-i}$	... $Z_{0,n-1}$	$Z_{0,n}$	
1	$Z_{1,0}$	$Z_{1,1}$	... $Z_{1,k}$	... $Z_{1,n-i}$	... $Z_{1,n-1}$	$Z_{1,n}$	
$\vdots$	$\vdots$	$\vdots$	$\vdots$	$\vdots$	$\vdots$	$\vdots$	
i	$Z_{i,0}$	$Z_{i,1}$	... $Z_{i,k}$	... $Z_{i,n-i}$	... $Z_{i,n-1}$	$Z_{i,n}$	
$\vdots$	$\vdots$	$\vdots$	$\vdots$	$\vdots$	$\vdots$	$\vdots$	
$n-k$	$Z_{n-k,0}$	$Z_{n-k,1}$	... $Z_{n-k,k}$	... $Z_{n-k,n-i}$	... $Z_{n-k,n-1}$	$Z_{n-k,n}$	
$\vdots$	$\vdots$	$\vdots$	$\vdots$	$\vdots$	$\vdots$	$\vdots$	
$n-1$	$Z_{n-1,0}$	$Z_{n-1,1}$	... $Z_{n-1,k}$	... $Z_{n-1,n-i}$	... $Z_{n-1,n-1}$	$Z_{n-1,n}$	
n	$Z_{n,0}$	$Z_{n,1}$	... $Z_{n,k}$	... $Z_{n,n-i}$	... $Z_{n,n-1}$	$Z_{n,n}$	

We assume that the incremental losses $Z_{i,k}$ are observable for $i + k \leq n$ and that they are non-observable for $i + k \geq n + 1$. For $i \in \{0, 1, \ldots, n\}$ we denote by

$$S_{i,n} := \sum_{l=0}^{n} Z_{i,l}$$

the ultimate loss of accident year i.

The Poisson model is a model for *claim numbers*:

Poisson Model:

(i) *The family of incremental losses* $\{Z_{i,k}\}_{i,k \in \{0,1,\ldots,n\}}$ *is independent.*

K.D. Schmidt (✉)
Technische Universität Dresden, Dresden, Germany
e-mail: klaus.d.schmidt@tu-dresden.de

© Springer International Publishing Switzerland 2016
M. Radtke et al. (eds.), *Handbook on Loss Reserving*,
EAA Series, DOI 10.1007/978-3-319-30056-6_30

(ii) *There exist parameters $\alpha_0, \alpha_1, \ldots, \alpha_n \in (0, \infty)$ and $\vartheta_0, \vartheta_1, \ldots, \vartheta_n \in (0, 1)$ with $\sum_{l=0}^{n} \vartheta_l = 1$ such that*

$$P_{Z_{i,k}} = \mathbf{Poi}(\alpha_i \vartheta_k)$$

holds for all $i, k \in \{0, 1, \ldots, n\}$.

In this article we assume that the assumptions of the Poisson model are fulfilled. Because of the identity

$$E[Z_{i,k}] = \alpha_i \, \vartheta_k$$

we are in the situation of a *multiplicative model*. Furthermore, summation yields

$$E[S_{i,n}] = \alpha_i$$

Therefore, the parameters $\alpha_0, \alpha_1, \ldots, \alpha_n$ are the expected ultimate losses of the accident years and the parameters $\vartheta_0, \vartheta_1, \ldots, \vartheta_n$ form a *development pattern for incremental quotas*.

To estimate the parameters, there are essentially two methods and they turn out to produce the same result.

Marginal Sum Method

Since the Poisson model is a multiplicative model for incremental losses, one method to estimate the parameters is the *marginal sum method* in which the *marginal sum estimators*

$$\alpha_0^{\mathrm{MS}}, \alpha_1^{\mathrm{MS}}, \ldots, \alpha_n^{\mathrm{MS}} \quad \text{and} \quad \vartheta_0^{\mathrm{MS}}, \vartheta_1^{\mathrm{MS}}, \ldots, \vartheta_n^{\mathrm{MS}}$$

are determined as a solution of the *marginal sum equations*

$$\sum_{l=0}^{n-i} \widehat{\alpha_i} \widehat{\vartheta_l} = \sum_{l=0}^{n-i} Z_{i,l}$$

with $i \in \{0, 1, \ldots, n\}$ and

$$\sum_{j=0}^{n-k} \widehat{\alpha_j} \widehat{\vartheta_k} = \sum_{j=0}^{n-k} Z_{j,k}$$

with $k \in \{0, 1, \ldots, n\}$ under the constraint

$$\sum_{l=0}^{n} \widehat{\vartheta_l} = 1$$

This system of non-linear equations has a unique solution and the marginal sum estimators of the expected ultimate losses are identical with the *chain ladder predictors* of the ultimate losses.

Maximum Likelihood Method

In the Poisson model the joint distribution of all incremental losses is known except for the parameters and it satisfies

$$P\left[\bigcap_{j=0}^{n}\bigcap_{l=0}^{n}\{Z_{j,l}=z_{j,l}\}\right]=\prod_{j=0}^{n}\prod_{l=0}^{n}\left(e^{-\alpha_j\vartheta_l}\frac{(\alpha_j\vartheta_l)^{z_{j,l}}}{z_{j,l}!}\right)$$

Therefore, the *maximum likelihood method*, which is based on the joint distribution of the observable incremental losses, can be used to estimate the parameters. Since

$$P\left[\bigcap_{j=0}^{n}\bigcap_{l=0}^{n-j}\{Z_{j,l}=z_{j,l}\}\right]=\prod_{j=0}^{n}\prod_{l=0}^{n-j}\left(e^{-\alpha_j\vartheta_l}\frac{(\alpha_j\vartheta_l)^{z_{j,l}}}{z_{j,l}!}\right)$$

the likelihood function L is given by

$$L(\widehat{\alpha}_0,\widehat{\alpha}_1,\ldots,\widehat{\alpha}_n,\widehat{\vartheta}_0,\widehat{\vartheta}_1,\ldots,\widehat{\vartheta}_n):=\prod_{j=0}^{n}\prod_{l=0}^{n-j}\left(e^{-\widehat{\alpha}_j\widehat{\vartheta}_l}\frac{(\widehat{\alpha}_j\widehat{\vartheta}_l)^{Z_{j,l}}}{Z_{j,l}!}\right)$$

The maximum likelihood method determines the *maximum likelihood estimators*

$$\alpha_0^{ML},\alpha_1^{ML},\ldots,\alpha_n^{ML}\quad\text{and}\quad\vartheta_0^{ML},\vartheta_1^{ML},\ldots,\vartheta_n^{ML}$$

as a common zero of the first order partial derivatives of the likelihood function L under the constraint

$$\sum_{l=0}^{n}\vartheta_l^{ML}=1$$

The common zeros of the partial derivatives of the likelihood function under the constraint are identical with those of the partial derivatives of the log-likelihood function $\ln(L)$ under the constraint, and they turn out to be the solutions of the equations

$$\sum_{l=0}^{n-i}\widehat{\alpha}_i\,\widehat{\vartheta}_l=\sum_{l=0}^{n-i}Z_{i,l}$$

with $i \in \{0, 1, \dots, n\}$ and

$$\sum_{j=0}^{n-k} \widehat{\alpha}_j \, \widehat{\vartheta}_k = \sum_{j=0}^{n-k} Z_{j,k}$$

with $k \in \{0, 1, \dots, n\}$ under the constraint

$$\sum_{l=0}^{n} \widehat{\vartheta}_l = 1$$

Therefore, the maximum likelihood estimators coincide with the marginal sum estimators.

Remarks

The following result is useful for the comparison of the Poisson model with other models:

Theorem. *The assumptions of the Poisson model are fulfilled if and only if the following conditions are fulfilled:*

(i) *The family* $\{\{Z_{i,k}\}_{k \in \{0,1,\dots,n\}}\}_{i \in \{0,1,\dots,n\}}$ *is independent.*

(ii) *There exist parameters* $\alpha_0, \alpha_1, \dots, \alpha_n \in (0, \infty)$ *such that*

$$P_{S_{i,n}} = \mathbf{Poi}(\alpha_i)$$

holds for each accident year $i \in \{0, 1, \dots, n\}$.

(iii) *There exist parameters* $\vartheta_0, \vartheta_1, \dots, \vartheta_n \in (0, 1)$ *with* $\sum_{l=0}^{n} \vartheta_l = 1$ *such that*

$$P_{Z_{i,0}, Z_{i,1}, \dots, Z_{i,n} | S_{i,n}} = \mathbf{M}(S_{i,n}; \vartheta_0, \vartheta_1, \dots, \vartheta_n)$$

holds for every accident year $i \in \{0, 1, \dots, n\}$.

Because of the theorem the Poisson model is a *multinomial model*. Furthermore, it fulfills the assumptions of the *credibility model of Witting*.

The first two conditions of the theorem are fulfilled when the ultimate losses are annual increments of an inhomogeneous Poisson process.

The Poisson model dates back to Hachemeister and Stanard [1975] and it is probably the first stochastic model which provides a justification of the chain ladder method by a classical statistical method of estimation.

Notes

Keywords: Chain Ladder Method (Basics), Credibility Models (Loss Reserving), Development Patterns (Basics), Marginal Sum Method, Multinomial Model, Multiplicative Models.

References: Hachemeister & Stanard [1975], Mack [2002], Schmidt [1996, 2009, 2012], Schmidt & Wünsche [1998].

Reinsurance

Michael Radtke

The use of actuarial methods to determine and evaluate loss reserves is common practice in reinsurance and belongs to the standard tasks of an actuarial department of a reinsurer.

Compared to a direct insurer, the same types of loss reserves appear in reinsurance, i.e. there are

- provisions for incurred and known individual losses (*individual loss reserves*),
- provisions for insufficiently reserved losses (IBNER reserves: *incurred but not enough reserved*) and
- provisions for incurred but not yet reported losses (reserves for late reported losses or IBNR reserves: *incurred but not reported*).

In addition, the direct and indirect *loss adjustment expenses* and the premium development in lines of business with premium run-off (the so-called *underwriting year lines of business*) have to be considered.

M. Radtke (✉)
Fachhochschule Dortmund, Dortmund, Germany
e-mail: michael.radtke@fh-dortmund.de

© Springer International Publishing Switzerland 2016
M. Radtke et al. (eds.), *Handbook on Loss Reserving*,
EAA Series, DOI 10.1007/978-3-319-30056-6_31

Consider the *run-off triangle* $\Delta(S)$ of the observable cumulative losses $S_{i,k}$ with $i + k \leq n$:

Accident year	Development year						
	0	1	... k	... $n-i$	... $n-1$	n	
0	$S_{0,0}$	$S_{0,1}$	... $S_{0,k}$	... $S_{0,n-i}$	... $S_{0,n-1}$	$S_{0,n}$	
1	$S_{1,0}$	$S_{1,1}$	... $S_{1,k}$	... $S_{1,n-i}$	... $S_{1,n-1}$		
$\vdots$	$\vdots$	$\vdots$	$\vdots$	$\vdots$			
i	$S_{i,0}$	$S_{i,1}$	... $S_{i,k}$	... $S_{i,n-i}$			
$\vdots$	$\vdots$	$\vdots$	$\vdots$				
$n-k$	$S_{n-k,0}$	$S_{n-k,1}$	... $S_{n-k,k}$				
$\vdots$	$\vdots$	$\vdots$					
$n-1$	$S_{n-1,0}$	$S_{n-1,1}$					
n	$S_{n,0}$						

In practice, the underlying data are often given in a different form and in different levels of aggregation and have to be transformed into a run-off triangle.

Besides the line of business, the type of reinsurance as well as the concrete parameters of the contract have a decisive influence on the actuarial methods of loss reserving to be applied.

Proportional Reinsurance

In *proportional reinsurance*, i.e. in quota share reinsurance and in surplus reinsurance, the development of reinsurance losses runs largely similar to that of the original losses.

A *quota share reinsurance* contract with a constant share $q \in (0, 1)$ transforms the cumulative loss $S_{i,k}$ of the original business (*paid loss* or *incurred loss* from accident year i until the end of development year k) into the cumulative loss $S_{i,k}^{R} := q\, S_{i,k}$ of the reinsurer and the original run-off triangle $\Delta(S)$ yields the run-off triangle

$$\Delta(S^{R}) := q\, \Delta(S)$$

of the reinsurer. It follows that

- the *development pattern*,
- the run-off triangle $\Delta(\varphi)$ of the *individual development factors*

$$\widehat{\varphi}_{i,k} := \frac{S_{i,k}}{S_{i,k-1}}$$

and
- the *chain ladder factors*

$$\widehat{\varphi}_k^{\mathrm{CL}} := \frac{\sum_{j=0}^{n-k} S_{j,k}}{\sum_{j=0}^{n-k} S_{j,k-1}} = \sum_{j=0}^{n-k} \frac{S_{j,k-1}}{\sum_{h=0}^{n-k} S_{h,k-1}} \, \widehat{\varphi}_{j,k}$$

remain unchanged at first. In practice, however, the share is usually not constant over the entire period under consideration. This effect can be adjusted by changing the weights if the shares for the different accident years are known.

A *surplus reinsurance* contract can be handled as a quota share reinsurance contract with segment specific shares q_m for the bands of sums insured S_m and with the run-off triangle

$$\Delta(S^R) := \sum q_m \, \Delta(S_m)$$

The development factors can then be estimated by a weighted average of the segment specific chain ladder factors.

Thus, in proportional reinsurance the same actuarial loss reserving approaches and techniques are employed as in direct insurance. The only difference for reinsurance is the usually longer time delay of the available run-off data. Sometimes additional problems arise from incomplete accounting data or from a mixture of regularly developing contracts and contracts with clean-cut or sunset agreements.

Non-proportional Reinsurance

By contrast, the situation in *non-proportional reinsurance* (NP) is completely different. In *excess of loss reinsurance* (XL) as well as in *stop loss reinsurance* (SL), the development pattern of the ceded cumulative losses is not comparable with that of the original business. In particular, in *long-tail* lines of business such as *industrial liability insurance*, *professional indemnity insurance* and *motor third party liability insurance* the share of IBNR losses for the reinsurer increases significantly.

Often no paid losses are observed at all in the first development years of an XL contract. Due to the increasing potential of large losses, the volatility of the run-off data increases significantly and therefore differentiated actuarial methods are required.

Further particularities have to be considered in reserving for NP contracts or entire NP portfolios, such as changing contract parameters like retentions, limits of liabilities, annual aggregate deductibles or index clauses. Frequently additional problems arise from incomplete accounting data or from a mixture of regularly developing contracts and contracts with commutation clauses, which have to be adequately adjusted at first.

Data

As a general rule, the data have to be prepared and validated before the application of actuarial loss reserving methods. Specific for reinsurance is that the reinsurance contract parameters may change from year to year. These changes often have a very decisive influence on the run-off behaviour.

In order to consider these effects reasonably, it is most suitable to register the *run-off data* on contract and loss level and to make appropriate adjustments before the usual run-off triangles are compiled. For example, run-off data with different retentions and limits can be compiled for an NP portfolio or the loss data can be inflated in case of an effective index clause.

In segments with hardly predictable business, like *hospital liability insurance*, additional run-off triangles for the *claim numbers* and the *average claim size* are required for a detailed run-off analysis.

Loss Reserving Methods in Reinsurance

A wide range of different actuarial methods is available for the calculation and evaluation of loss reserves. The basic methods that are applied in reinsurance are

- the *additive method*,
- the *chain ladder method*,
- the *loss development method*,
- the *Bornhuetter–Ferguson method* and
- the *Cape Cod method*.

The assumption of a *development pattern* is common to all these methods. Since the estimators of the development factors are sometimes called *loss development factors* (LDF), these methods are also referred to as *LDF-based methods*. In practice modifications and extensions of these methods are often used.

For a deeper analysis *credibility models* or *generalized linear models* can be employed, whereas in the case of missing run-off data very simple methods such as the *expected loss method* may have to be used.

In the following sections we discuss some reinsurance specific aspects.

Basic Model of LDF-Based Methods

Many actuarial methods of loss reserving are based on the assumption that a development pattern for factors exists. Therefore, the analysis and the estimation of the development factors is an important process step after the preparation of a comprehensive data basis.

In a basic model for this class of methods we assume that the conditional expectations of the cumulative losses fulfill the condition

$$E\left(S_{i,k} \mid S_{i,0}, S_{i,1}, \ldots, S_{i,k-1}\right) = S_{i,k-1}\,\varphi_k$$

with a *development factor* φ_k which is independent of the accident year i. If we assume further that the conditional variances fulfill the condition

$$\mathrm{var}\left(S_{i,k} \mid S_{i,0}, S_{i,1}, \ldots, S_{i,k-1}\right) = S_{i,k-1}\,\sigma_k^2$$

with a parameter σ_k^2 which is independent of the accident year i and that the accident years are independent, then we obtain the *chain ladder model of Mack*. This model can be generalized to the *chain ladder model of Schnaus* and the optimal estimators of the development factors are the chain ladder factors

$$\varphi_k^{\mathrm{CL}} := \sum_{j=0}^{n-k} W_{j,k}^{\mathrm{CL}}\,\widehat{\varphi}_{j,k}$$

with the individual development factors

$$\widehat{\varphi}_{i,k} := \frac{S_{i,k}}{S_{i,k-1}}$$

and the *chain ladder weights*

$$W_{i,k}^{\mathrm{CL}} := \frac{S_{i,k-1}}{\sum_{h=0}^{n-k} S_{h,k-1}}$$

In this stochastic model the standard errors of the chain ladder reserves can be estimated and, by making additional distributional assumptions, also confidence intervals can be determined.

Methodological Adjustments

In reinsurance, and especially in non-proportional reinsurance, it is, due to structural breaks, trends or insufficient data, quite often necessary to make methodological adjustments and modifications of the standard methods.

In the following we present some typical problems and approaches for their solution.

Restriction of the Observation Period

In case of a systematic break in the run-off data, which may be caused by a substantial change of the reinsurance structure or by highly erroneous data, a reduction of the run-off data to a subset including, for example, only the younger accident or calendar years, may be necessary and appropriate.

Smoothing the Loss Development Factors

The loss development factors can be considered as the values of a function of the development years and can be modelled by a classical regression approach. Correspondingly, a model like $E[\varphi_k^{CL}] = \varphi(k)$ with an appropriate function φ can be considered for the expectations of the chain ladder factors. Such smoothing approaches are, in particular, made for NP reinsurance contracts and for portfolios with thin data and high volatility.

Tail

If even the oldest available accident year is not yet completely developed, *tail factors* are required, which specify the further development to the end. The tail can be obtained either by extrapolation of the loss development factors or from generally accessible market information. If in addition the interest rate effect should be considered as well and loss reserves should be discounted, then the *tail pattern* has to be modelled as a *cash flow pattern*.

Weighting and Trending of the Individual Development Factors

Another analytical step is the examination of the individual development factors. Often a trend over the accident years can be observed for a fixed development year. A trend over the accident years can be caused by

- a changing portfolio composition,
- changing contract parameters of the underlying reinsurance contracts or
- a changing claims handling practice.

Again, we can obtain estimators of the development factors by an appropriate regression approach for the individual development factors. Another common method is the weighting of the individual development factors with weights depending on the accident years. The chain ladder factors

$$\varphi_k^{\mathrm{CL}} := \frac{\sum_{j=0}^{n-k} S_{j,k}}{\sum_{j=0}^{n-k} S_{j,k-1}} = \sum_{j=0}^{n-k} \frac{S_{j,k-1}}{\sum_{h=0}^{n-k} S_{h,k-1}} \, \widehat{\varphi}_{j,k}$$

already use such weights. To reinforce the dependence of the weights on the accident years, the chain ladder weights may be modified by using known volume measures $v_0, v_1, \ldots, v_n$ of the accident years to obtain the *modified chain ladder factors*

$$\widehat{\varphi}_k := \frac{\sum_{j=0}^{n-k} v_j \, S_{j,k}}{\sum_{j=0}^{n-k} v_j \, S_{j,k-1}} = \sum_{j=0}^{n-k} \frac{v_j \, S_{j,k-1}}{\sum_{h=0}^{n-k} v_h \, S_{h,k-1}} \, \widehat{\varphi}_{j,k}$$

Another possibility, which is frequently used, is to assign higher weights to the individual development factors of younger accident years and lower weights to those of older ones. Finally, the introduction of a credibility model is a more advanced approach, in particular if external estimators of the development factors or estimators from upper aggregation levels are to be used in addition, such as estimators of the corresponding portfolio besides the contract specific estimators for example.

Adjustment of Large Losses

Large losses as well as a systematical absence of losses lead to strong distortions in the development behaviour and in the estimation of the development variables, in particular if they occur in the current business year (hence on the main diagonal of the run-off triangle). Most LDF-based methods are, due to their multiplicative structure, very sensitive to such *outliers*. For example, the chain ladder method, and also the loss development method, projects such an effect proportionally onto the predictor of the ultimate loss. Therefore, either an individual adjustment is necessary before the application of these methods or a less sensitive method has to be used. The *Cape Cod method* is such a method since it adjusts distorted current losses automatically. Its basic idea is the splitting

$$S_{i,n-i} = T_{i,n-i} + X_i$$

of the current losses $S_{i,n-i}$ into the *smoothed cumulative losses* $T_{i,n-i}$, which are developed by the *loss development method*, and the *outlier effects* X_i, which are carried along additively without any development. Another possibility is the application of the *Bornhuetter–Ferguson method* with appropriate a priori estimators of the quotas and of the expected ultimate losses, which could be obtained from external data for example.

Inflation Adjustment

In long-tail business, inflation has to be handled separately if it is significant. The standard methods, like the chain ladder method, do not take inflation effects appropriately into account, in particular, if the inflation is strongly fluctuating or shows a trend over the calendar years. A possibility is the calculation of inflation adjusted run-off triangles. On the basis of the original incremental losses $Z_{i,k} := S_{i,k} - S_{i,k-1}$ the run-off triangle of the present values is calculated in order to apply a classical loss reserving method to it. Subsequently the predictors of the future cumulative losses have to be inflated with appropriate inflation rates. Alternatively, some reserving methods like the *separation method* may be used to model inflation effects directly with calendar year specific factors.

Adaption of the A Priori Estimators in the Bornhuetter–Ferguson Method

The *Bornhuetter–Ferguson predictors*

$$S_{i,k}^{\mathrm{BF}} := S_{i,n-i} + \left(\widehat{\gamma}_k - \widehat{\gamma}_{n-i}\right)\widehat{\alpha}_i$$

involve *a priori estimators* $\widehat{\gamma}_0, \widehat{\gamma}_1, \ldots, \widehat{\gamma}_n$ with $\widehat{\gamma}_n := 1$ of the *quotas* $\gamma_0, \gamma_1, \ldots, \gamma_n$ and *a priori estimators* $\widehat{\alpha}_0, \widehat{\alpha}_1, \ldots, \widehat{\alpha}_n$ of the *expected ultimate losses* $\alpha_0, \alpha_1, \ldots, \alpha_n$. Typically the expected ultimate losses are represented in the form

$$\alpha_i = v_i \kappa_i$$

where v_i is a *volume measure* and κ_i is the corresponding *expected ultimate loss ratio* of accident year i. The a priori estimators of the expected ultimate losses then result from a priori estimators $\widehat{\kappa}_0, \widehat{\kappa}_1, \ldots, \widehat{\kappa}_n$ of the expected ultimate loss ratios by letting

$$\widehat{\alpha}_i := v_i \widehat{\kappa}_i$$

Besides the choice of the a priori estimators of the quotas, the choice of the a priori estimators of the expected ultimate losses is decisive. For reinsurance portfolios it is often difficult to obtain reliable a priori estimators of the expected ultimate losses of the younger accident years. In such cases an iteration of the Bornhuetter–Ferguson method can make sense. In doing so, the method is applied a second time, now replacing the a priori estimators $\widehat{\alpha}_i$ by the Bornhuetter–Ferguson predictors $S_{i,n}^{\mathrm{BF}}$ obtained from the first application. This yields the *Benktander–Hovinen predictors*.

Application Areas in Reinsurance

Actuarial methods of loss reserving are applied in numerous reinsurance situations. In long-tail lines of business such as *industrial liability insurance, professional indemnity insurance*, and *motor third party liability insurance* the pricing of reinsurance contracts is substantially based on the development of future losses. For the pricing of a reinsurance contract, an actuarial prediction of the ultimate loss and an accurate quantification of the development pattern are essential for the determination of the present value of the future payment obligations.

Another area of application of actuarial reserve analyses is the risk and portfolio management of a reinsurer. The modelling and measurement of the development is a central component of every *controlling approach* of a reinsurer. Thus the quantification of the expected run-off profit or loss of an operational business unit of a reinsurer is indispensable for the evaluation of the profitability of this unit.

However, actuarial methods are not only applied in the internal determination of reserve requirements and the evaluation of reserve quality, but actuarial loss reserving methods also represent an important toolkit for commutations and *portfolio transfers* between insurance companies or for the evaluation of entire insurance companies.

Finally the loss reserves usually establish the largest item on the liabilities side of a reinsurer's balance sheet. Therefore in the *financial accounting* framework high requirements on the well-founded disclosure and particularly on the adequacy are set. Supervisory rules and international financial accounting standards often demand the application of actuarial loss reserving methods and an actuarial certification by an appointed actuary.

Notes

Keywords: Additive Method, Bornhuetter–Ferguson Method, Bornhuetter–Ferguson Principle, Cape Cod Method, Chain Ladder Method (Basics), Chain Ladder Method (Models), Controlling, Credibility Models (Loss Reserving), Development Patterns (Basics), Development Patterns (Estimation), Expected Loss Method, Loss Development Method, Run-Off Data, Run-Off Triangles, Separation Method, Tail.

Literature: Benktander [1976], Mack [1993, 2002], Neuhaus [1992], Radtke [2008], Schmidt [2009], Schmidt & Schnaus [1996], Schweizer Rück [1989], Straub [1988], Taylor [1986, 2000], Van Eeghen [1981].

Run-Off Data

Michael Radtke

For the application of actuarial reserving methods, the availability of a valid and comprehensive data basis is an essential precondition and constitutes a vital success factor. The compilation of a sound data basis usually presents a substantial part of the effort during a run-off analysis.

In the development of a data concept and the subsequent construction of a data warehouse, numerous aspects have to be considered. These include

- the data organization,
- the data definition and
- the data preparation,

which will be discussed in this article.

Data Organization

Actuarial reserving methods aim at the modelling and the prediction of the development of losses in several temporal dimensions. The underlying data should appropriately represent the temporal dimensions in terms of process data, and the loss development is understood as a stochastic process of the target quantities under consideration.

M. Radtke (✉)
Fachhochschule Dortmund, Dortmund, Germany
e-mail: michael.radtke@fh-dortmund.de

© Springer International Publishing Switzerland 2016
M. Radtke et al. (eds.), *Handbook on Loss Reserving*,
EAA Series, DOI 10.1007/978-3-319-30056-6_32

In a run-off analysis, the temporal dimensions of the target quantities under consideration are

- the *accident year* (or the *underwriting year*),
- the *development year* and
- the *calendar year* (or the *accounting year*).

These temporal dimensions can be represented in a *run-off triangle*.

Example. We denote by $S_{i,k}$ the sum of all payments for losses from the relative accident year i that are made before the end of the relative development year k (and hence before the end of the relative calendar year $i + k$). We obtain the following run-off triangle of cumulative losses:

Accident	Development year							
year	0	1	... k		... $n-i$	... $n-1$	n
0	$S_{0,0}$	$S_{0,1}$	... $S_{0,k}$		... $S_{0,n-i}$	... $S_{0,n-1}$	$S_{0,n}$
1	$S_{1,0}$	$S_{1,1}$	... $S_{1,k}$		... $S_{1,n-i}$	... $S_{1,n-1}$	
$\vdots$	$\vdots$	$\vdots$	$\vdots$	$\vdots$			
i	$S_{i,0}$	$S_{i,1}$	... $S_{i,k}$		... $S_{i,n-i}$		
$\vdots$	$\vdots$	$\vdots$	$\vdots$				
$n-k$	$S_{n-k,0}$	$S_{n-k,1}$	... $S_{n-k,k}$				
$\vdots$	$\vdots$	$\vdots$					
$n-1$	$S_{n-1,0}$	$S_{n-1,1}$					
n	$S_{n,0}$						

Since the sum of the indices of the cumulative losses is constant on every diagonal, all cumulative losses on a diagonal belong to the same calendar year.

In practice the run-off data are not always available in this structured form, but are provided as sequential aggregated data records or as single data records by an operative system.

Example. This example illustrates the development of three contracts in the first three development years. The yearly change is shown for all values. For example, for the first contract a payment of 2934 is made in the year 2000 and a reserve of 2176 is established. In the year 2001, a further payment of 1882 is made for this contract and the reserve is reduced by 1754.

Ins. no.	Pol. no.	LoB	Cover	Acc. year	Dev. year	Earned premium	Paid costs	Reserve loss	Acc. change	Status
10001	4001	F	3	2000	2000	4592	−1453	−2934	−2176	1
10001	4001	F	3	2000	2001	0	2	−1882	1754	1
10001	4001	F	3	2000	2002	0	0	0	0	1
10001	4001	F	3	2001	2001	4983	−1770	−117	−2232	1
10001	4001	F	3	2001	2002	0	3	−1169	1556	1
10002	4002	F	3	2000	2000	403	−123	−13	−156	1
10002	4002	F	3	2000	2001	− 1	0	−45	−232	1
10002	4002	F	3	2000	2002	0	0	0	0	1
10002	4002	F	3	2001	2001	405	−123	−16	− 71	1
10002	4002	F	3	2001	2002	0	0	0	71	1
10002	4003	F	3	2000	2000	2244	−1116	−251	−987	1
10002	4003	F	3	2000	2001	−34	0	−791	953	1
10002	4003	F	3	2000	2002	0	0	0	0	1
10002	4003	F	3	2001	2001	2359	−1209	−250	−837	1
10002	4003	F	3	2001	2002	0	0	0	0	1

Here the single loss data records are already aggregated at contract level and at a valuation date at the end of the development year.

The access to contract and loss data records is helpful for the validation of the run-off data. The influence of large losses on the loss development can be investigated in detail. Moreover, the effects of reinsurance can be analysed comprehensively only on this level, and gross and net run-off triangles can be constructed, especially for *non-proportional reinsurance*.

Data Definition

In a run-off analysis, several target quantities can be recorded in their temporal development and they can be analysed with actuarial methods.

The following basic quantities are considered in any run-off analysis:

- the *paid losses* (*paid* for short);
- the *loss reserves* (*reserve* for short);
- the resulting *incurred losses* (*incurred* for short).

In addition, *premiums* and *costs* are usually analysed as well so that an economic assessment can be conducted.

The precise definition of these target quantities is extremely important. Income and expense items have to be clearly distinguished and defined in their temporal dimension and in their status. For instance, it has to be clarified,

- if the premium is written, accounted or earned,
- if the paid claims include salvages and subrogations or
- if the *individual loss reserves* are established with *lump sum approaches* in certain circumstances.

Moreover, other appropriate aspects should be considered as well in specific applications.

Comprehensive and detailed analyses also include the *number of claims* and the *average claim size*. For this purpose, the number of risks or contracts is required besides the number of claims or, in general, a *volume measure* such as the sum insured or the limit of liability or the premium calculation basis in *liability insurance*.

Example. For *hospital liability insurance*, this means that, besides the premium information, the number of beds and the number of claims be available such that the development of the incurred losses per hospital bed and per claim can be estimated within the run-off analysis.

Another topic is the examination of the development of *relative quantities*. The most common *standardizations* use

- premiums,
- estimated ultimate losses or
- paid losses.

The following table presents some possible relative quantities:

	Premium	Ultimate incurred loss	Paid loss
	P	E	Z
Paid loss	paid loss ratio	paid quota	
Z	Z/P	Z/E	
Incurred loss	incurred loss ratio	incurred quota	incurred/paid
A	A/P	A/E	relation A/Z
Loss reserve	reserve loss ratio	reserved quota	reserve/paid
R	R/P	R/E	relation R/Z

Run-off triangles can be compiled for each of these relative quantities.

Some of these relative quantities have an interesting counterpart referring to aggregated basic quantities: For example, the paid quotas constructed from paid losses and ultimate incurred losses, both aggregated over all observable accident years, yield an empirical *development pattern* for paid losses, which can be represented as a *development curve*.

Data Preparation

An essential precondition for the application of actuarial analysis methods is the *homogeneity* of the underlying basic data. The homogeneity has to be ensured by an appropriate *segmentation* during the data preparation. Beyond the distinction of different lines of business, the segments are defined in such a way that the assumption of a uniform development of the individual losses is likely to be justified. Commonly accepted criteria for segmentation are lines and sub-lines of business, business units, regions and channels of distribution, and in reinsurance also the type of the reinsurance treaty. However, excessive segmentation with small segment volumes may result in a very volatile run-off behaviour, which causes problems in the application of actuarial methods.

Data validation is another important aspect of data preparation. If the run-off data are available on the basis of single data records, then *large losses* or losses with an unusual development can be identified and treated separately and one may decide whether and in which way they should be included in the subsequent actuarial analysis. For example, single large losses could be capped; also, when available, statistics of large claims for similar segments or the market might be helpful to find an appropriate treatment of large claims. By contrast, the systematic elimination of large claims should be avoided.

In general, the comparison of the individual run-off data and the results of the actuarial analysis with general available market data in the terms of a *benchmarking* represents another step in quality management.

Notes

Keywords: Aggregation, Development Patterns (Basics), Reinsurance, Run-Off Triangles, Volume Measures.

Run-Off Triangles

Klaus D. Schmidt

In this article we assume that the *accounting years* match the *calendar years*. For simplification we speak about *years*.

Every claim has a history:

- The claim occurs in an *accident year*.
- The claim is reported to the insurance company.
- The insurance company provides first payments and establishes an *individual loss reserve* for potential further payments.
- The claim is finally settled.

The settlement of an individual claim and hence the settlement of all claims of a given accident year can extend over several *development years*.

The *run-off data* that are accrued during the development of a portfolio can be structured in different aspects in form and content. With regard to the determination of loss reserves, it is most convenient to represent the run-off data by accident and development years in the form of a *run-off triangle*.

A run-off triangle can be used to represent, for every accident year and every development year,

- the *number of claims*, i.e. the number of reported claims,
- the *paid losses*,
- the *incurred losses*, i.e. the sum of the paid losses and the individual loss reserves,

or even some other quantities which are of interest in loss reserving.

In order to adequately represent the complexity of the run-off data, it is sometimes necessary to consider simultaneously several run-off triangles for different types of data.

K.D. Schmidt (✉)
Technische Universität Dresden, Dresden, Germany
e-mail: klaus.d.schmidt@tu-dresden.de

© Springer International Publishing Switzerland 2016
M. Radtke et al. (eds.), *Handbook on Loss Reserving*,
EAA Series, DOI 10.1007/978-3-319-30056-6_33

For the sake of simplification, the data of a run-off triangle will be interpreted as paid losses in the present article.

Example. The following table contains the paid losses for claims from accident years 2010 to 2015 that have been paid in the individual development years before the end of year 2015:

Accident	Development year					
year	2010	2011	2012	2013	2014	2015
2010	1001	854	568	565	347	148
2011		1113	990	671	648	422
2012			1265	1168	800	744
2013				1490	1383	1007
2014					1725	1536
2015						1889

For claims from accident year 2013, the amount of 1490 has been paid in the same year and the amounts of 1383 respectively 1007 have been paid in the subsequent development years.

The representation of run-off data becomes more meaningful by passing to a *run-off triangle* in which the development years are not noted as calendar years but as delays with respect to the accident years and hence as *relative development years*:

Example. With the transition to relative development years we obtain the following run-off triangle:

Accident	Development year					
year	0	1	2	3	4	5
2010	1001	854	568	565	347	148
2011	1113	990	671	648	422	
2012	1265	1168	800	744		
2013	1490	1383	1007			
2014	1725	1536				
2015	1889					

The amounts paid in 2015 for claims from accident years 2010 to 2015 appear on the main diagonal.

The run-off triangle allows us to detect whether the paid losses show a *trend* over the accident years or follow a *development pattern* over the development years which is common to all accident years. From the trend and the development pattern it is then possible to get a rough idea of the amount of future payments and hence of the necessary *loss reserve*.

Once the transition to relative development years has been done, it is then quite natural to proceed in the same way with the accident years and to note them as *relative accident years.* This notation will also turn out to be most convenient in the mathematical formulation of models and methods in loss reserving.

Example. With the transition to relative accident years we obtain the following run-off triangle:

Accident	Development year					
year	0	1	2	3	4	5
0	1001	854	568	565	347	148
1	1113	990	671	648	422	
2	1265	1168	800	744		
3	1490	1383	1007			
4	1725	1536				
5	1889					

This run-off triangle contains the paid losses in the relative development years for the relative accident years.

The last run-off triangle provides the basis for all further considerations.

Run-Off Squares and Run-Off Triangles of Random Variables

We consider $n + 1$ accident years and we assume that every claim is finally settled either in the accident year or in one of the subsequent n calender years. We sometimes refer to n as the *development horizon.*

We consider further a family $\{Z_{i,k}\}_{i,k\in\{0,1,...,n\}}$ of random variables and we interpret $Z_{i,k}$ as the payment in the (relative) development year k for the claims from the (relative) accident year i. The payment $Z_{i,k}$ is then made in the *(relative) calendar year* $i + k$. The random variables $Z_{i,k}$ are called *incremental losses* and we assume that the incremental losses are observable (but not yet observed) for $i + k \leq n$ and that they are non-observable for $i + k \geq n + 1$.

The incremental losses are represented in a *run-off square of incremental losses:* Then the observable incremental losses form a *run-off triangle of incremental losses:* The sum of the indices of the random variables on the main diagonal is always equal to n, which is also referred to as the *current calendar year.*

Example. The run-off triangle

is a realization of the run-off triangle of incremental losses with $n = 5$. Besides the incremental losses $Z_{i,k}$ we also consider the *cumulative losses*

$$S_{i,k} := \sum_{l=0}^{k} Z_{i,l}$$

Then we have

Accident year	Development year					
	0 1 ... k ... $n-i$ ... $n-1$ n					
0	$Z_{0,0}$ $Z_{0,1}$ ... $Z_{0,k}$ ... $Z_{0,n-i}$ ... $Z_{0,n-1}$ $Z_{0,n}$					
1	$Z_{1,0}$ $Z_{1,1}$ ... $Z_{1,k}$ ... $Z_{1,n-i}$ ... $Z_{1,n-1}$ $Z_{1,n}$					
i	$Z_{i,0}$ $Z_{i,1}$ ... $Z_{i,k}$ ... $Z_{i,n-i}$ ... $Z_{i,n-1}$ $Z_{i,n}$					
$n-k$	$Z_{n-k,0}$ $Z_{n-k,1}$ ... $Z_{n-k,k}$ ... $Z_{n-k,n-i}$ ... $Z_{n-k,n-1}$ $Z_{n-k,n}$					
$n-1$	$Z_{n-1,0}$ $Z_{n-1,1}$ ... $Z_{n-1,k}$ ... $Z_{n-1,n-i}$ ... $Z_{n-1,n-1}$ $Z_{n-1,n}$					
n	$Z_{n,0}$ $Z_{n,1}$ ... $Z_{n,k}$ ... $Z_{n,n-i}$ ... $Z_{n,n-1}$ $Z_{n,n}$					

Accident year	Development year					
	0 1 ... k ... $n-i$ ... $n-1$ n					
0	$Z_{0,0}$ $Z_{0,1}$ ... $Z_{0,k}$ ... $Z_{0,n-i}$ ... $Z_{0,n-1}$ $Z_{0,n}$					
1	$Z_{1,0}$ $Z_{1,1}$ ... $Z_{1,k}$ ... $Z_{1,n-i}$ ... $Z_{1,n-1}$					
i	$Z_{i,0}$ $Z_{i,1}$ ... $Z_{i,k}$ ... $Z_{i,n-i}$					
$n-k$	$Z_{n-k,0}$ $Z_{n-k,1}$ ... $Z_{n-k,k}$					
$n-1$	$Z_{n-1,0}$ $Z_{n-1,1}$					
n	$Z_{n,0}$					

Accident year	Development year					
	0	1	2	3	4	5
0	1001	854	568	565	347	148
1	1113	990	671	648	422	
2	1265	1168	800	744		
3	1490	1383	1007			
4	1725	1536				
5	1889					

$$Z_{i,k} = \begin{cases} S_{i,0} & \text{if } k = 0 \\ S_{i,k} - S_{i,k-1} & \text{else} \end{cases}$$

We interpret $S_{i,k}$ as the sum of the payments made in development years $l \in \{0, 1, \ldots, k\}$ for the claims from accident year i. These payments are made in the calendar years $p \in \{i, i+1, \ldots, i+k\}$.

According to our assumption on the incremental losses, the cumulative losses $S_{i,k}$ are observable for $i + k \leq n$ and they are non-observable for $i + k \geq n + 1$.

The cumulative losses are represented in a *run-off square of cumulative losses*:

Accident year	Development year						
	0	1	... k	... $n-i$	... $n-1$	n	
0	$S_{0,0}$	$S_{0,1}$	... $S_{0,k}$	... $S_{0,n-i}$	... $S_{0,n-1}$	$S_{0,n}$	
1	$S_{1,0}$	$S_{1,1}$	... $S_{1,k}$	... $S_{1,n-i}$	... $S_{1,n-1}$	$S_{1,n}$	
$\vdots$	$\vdots$	$\vdots$	$\vdots$	$\vdots$	$\vdots$	$\vdots$	
i	$S_{i,0}$	$S_{i,1}$	... $S_{i,k}$	... $S_{i,n-i}$	... $S_{i,n-1}$	$S_{i,n}$	
$\vdots$	$\vdots$	$\vdots$	$\vdots$	$\vdots$	$\vdots$	$\vdots$	
$n-k$	$S_{n-k,0}$	$S_{n-k,1}$	... $S_{n-k,k}$	... $S_{n-k,n-i}$	... $S_{n-k,n-1}$	$S_{n-k,n}$	
$\vdots$	$\vdots$	$\vdots$	$\vdots$	$\vdots$	$\vdots$	$\vdots$	
$n-1$	$S_{n-1,0}$	$S_{n-1,1}$	... $S_{n-1,k}$	... $S_{n-1,n-i}$	... $S_{n-1,n-1}$	$S_{n-1,n}$	
n	$S_{n,0}$	$S_{n,1}$	... $S_{n,k}$	... $S_{n,n-i}$	... $S_{n,n-1}$	$S_{n,n}$	

The observable cumulative losses then form a *run-off triangle of cumulative losses*:

Accident year	Development year						
	0	1	... k	... $n-i$	... $n-1$	n	
0	$S_{0,0}$	$S_{0,1}$	... $S_{0,k}$	... $S_{0,n-i}$	... $S_{0,n-1}$	$S_{0,n}$	
1	$S_{1,0}$	$S_{1,1}$	... $S_{1,k}$	... $S_{1,n-i}$	... $S_{1,n-1}$		
$\vdots$	$\vdots$	$\vdots$	$\vdots$	$\vdots$			
i	$S_{i,0}$	$S_{i,1}$	... $S_{i,k}$	... $S_{i,n-i}$			
$\vdots$	$\vdots$	$\vdots$	$\vdots$				
$n-k$	$S_{n-k,0}$	$S_{n-k,1}$	... $S_{n-k,k}$				
$\vdots$	$\vdots$	$\vdots$					
$n-1$	$S_{n-1,0}$	$S_{n-1,1}$					
n	$S_{n,0}$						

We refer to

$$S_{i,n-i}$$

as the *current loss* and to

$$S_{i,n}$$

as the *ultimate loss* of accident year i.

Example. The run-off triangle

Accident	Development year					
year	0	1	2	3	4	5
0	1001	1855	2423	2988	3335	3483
1	1113	2103	2774	3422	3844	
2	1265	2433	3233	3977		
3	1490	2873	3880			
4	1725	3261				
5	1889					

is a realization of the run-off triangle of cumulative losses with $n = 5$.

Reserves

The sum of all non-observable incremental losses

$$R := \sum_{l=1}^{n} \sum_{j=n-l+1}^{n} Z_{j,l} = \sum_{j=1}^{n} \sum_{l=n-j+1}^{n} Z_{j,l}$$

is called the (required) *aggregate loss reserve*. The aggregate loss reserve can be split by accident years or by calendar years:

- For $i \in \{1, \ldots, n\}$, the sum

$$R_i := \sum_{l=n-i+1}^{n} Z_{i,l}$$

is called the (required) *accident year reserve* for accident year i.
- For $c \in \{n + 1, \ldots, 2n\}$, the sum

$$R_{(c)} := \sum_{l=c-n}^{n} Z_{c-l,l}$$

is called the (required) *calendar year reserve* for the calendar year c.

Then we have

$$R = \sum_{i=1}^{n} R_i = \sum_{c=n+1}^{2n} R_{(c)}$$

Since the reserves are non-observable, *predictors* of the reserves have to be determined.

Since every reserve is a sum of non-observable incremental losses, predictors of all reserves can be obtained in a systematic way as follows: First, for every non-observable incremental loss $Z_{i,k}$, construct an observable random variable

$$\widehat{Z}_{i,k}$$

which serves as a predictor of that particular incremental loss, and then use the sums

$$\widehat{R} := \sum_{l=1}^{n} \sum_{j=n-l+1}^{n} \widehat{Z}_{j,l}$$

$$\widehat{R}_{i} := \sum_{l=n-i+1}^{n} \widehat{Z}_{i,l}$$

$$\widehat{R}_{(c)} := \sum_{l=c-n}^{n} \widehat{Z}_{c-l,l}$$

as predictors of the corresponding reserves.

Predictors of non-observable incremental losses can in turn be obtained by constructing first, for every non-observable cumulative loss $S_{i,k}$, an observable random variable

$$\widehat{S}_{i,k}$$

which serves as a predictor of that particular cumulative loss, and then using the differences

$$\widehat{Z}_{i,k} := \begin{cases} \widehat{S}_{i,n-i+1} - S_{i,n-i} & \text{if } k = n-i+1 \\ \widehat{S}_{i,k} - \widehat{S}_{i,k-1} & \text{else} \end{cases}$$

as predictors of the non-observable incremental losses $Z_{i,k}$.

Conversely, if the predictors of the non-observable incremental losses are given, then letting

$$\widehat{S}_{i,n} := S_{i,n-i} + \sum_{l=n-i+1}^{n} \widehat{Z}_{i,l}$$

yields predictors of the non-observable cumulative losses.

In any case, the predictors of the accident year reserves satisfy

$$\widehat{R}_{i} = \widehat{S}_{i,n} - S_{i,n-i}$$

This means that the predictors of the accident year reserves, and hence the predictor of the aggregate loss reserve, can be obtained directly from the predictors of the ultimate losses.

Estimators and Predictors

The observable random variables which are used to predict non-observable incremental or cumulative losses are either estimators or predictors according to the point of view.

Quite generally, an *estimator* is an observable random variable which serves as a substitute for an unknown parameter of the distribution of a random variable, whereas a *predictor* is an observable random variable which serves as a substitute for a non-observable random variable.

Therefore, an observable random variable $\widehat{Z}_{i,k}$ can be considered either as an estimator of the *expected future incremental loss* $E[Z_{i,k}]$ or as a predictor of the future incremental loss $Z_{i,k}$ with $i + k \geq n + 1$.

Accordingly, an observable random variable $\widehat{S}_{i,k}$ can be considered either as an estimator of the *expected future cumulative loss* $E[S_{i,k}]$ or as a predictor of the future cumulative loss $S_{i,k}$ with $i + k \geq n + 1$.

In particular, any estimator of the *expected ultimate loss* $E[S_{i,n}]$ is at the same time a predictor of the ultimate loss $S_{i,n}$, and vice versa.

Estimation Error and Prediction Error

Besides estimation of an unknown parameter, one is also interested in the accuracy of its estimators, and the usual quantity to measure the quality of an estimator is the *expected squared estimation error*.

Accordingly, besides prediction of a non-observable random variable, one is also interested in the accuracy of its predictors, and the usual quantity to measure the quality of a predictor is the *expected squared prediction error*.

In general, the expected squared estimation error and the expected squared prediction error have to be estimated as well.

Notes

Keywords: Development Patterns (Basics), Run-Off Data.

Separation Method

Michael Radtke

The development of losses can be examined in different temporal dimensions. The most familiar view of actuarial models in loss reserving involves only the *accident year* and the *development year*, and hence two of these dimensions. Another dimension is the *calendar year*. The run-off data of a calendar year are located on a diagonal of a run-off triangle or a run-off square.

Besides the familiar factors reflecting the influence of the accident years or the development years, the separation method also incorporates factors which reflect the influence of the calendar years, such as *inflation* or accounting year specific changes in the development of losses.

Consider the run-off square of incremental losses:

Accident year	Development year						
	0	1	... k	... $n-i$	... $n-1$	n
0	$Z_{0,0}$	$Z_{0,1}$	... $Z_{0,k}$	... $Z_{0,n-i}$	... $Z_{0,n-1}$	$Z_{0,n}$
1	$Z_{1,0}$	$Z_{1,1}$	... $Z_{1,k}$	... $Z_{1,n-i}$	... $Z_{1,n-1}$	$Z_{1,n}$
$\vdots$	$\vdots$	$\vdots$	$\vdots$	$\vdots$	$\vdots$	$\vdots$
i	$Z_{i,0}$	$Z_{i,1}$	... $Z_{i,k}$	... $Z_{i,n-i}$	... $Z_{i,n-1}$	$Z_{i,n}$
$\vdots$	$\vdots$	$\vdots$	$\vdots$	$\vdots$	$\vdots$	$\vdots$
$n-k$	$Z_{n-k,0}$	$Z_{n-k,1}$	... $Z_{n-k,k}$	... $Z_{n-k,n-i}$	... $Z_{n-k,n-1}$	$Z_{n-k,n}$
$\vdots$	$\vdots$	$\vdots$	$\vdots$	$\vdots$	$\vdots$	$\vdots$
$n-1$	$Z_{n-1,0}$	$Z_{n-1,1}$	... $Z_{n-1,k}$	... $Z_{n-1,n-i}$	... $Z_{n-1,n-1}$	$Z_{n-1,n}$
n	$Z_{n,0}$	$Z_{n,1}$	... $Z_{n,k}$	... $Z_{n,n-i}$	... $Z_{n,n-1}$	$Z_{n,n}$

M. Radtke (✉)
Fachhochschule Dortmund, Dortmund, Germany
e-mail: michael.radtke@fh-dortmund.de

© Springer International Publishing Switzerland 2016
M. Radtke et al. (eds.), *Handbook on Loss Reserving*,
EAA Series, DOI 10.1007/978-3-319-30056-6_34

We assume that the incremental losses $Z_{i,k}$ are observable for $i + k \leq n$ and that they are non-observable for $i + k \geq n + 1$.

The separation method is based on the following separation model and models besides the accident year effects and the development year effects also the calendar year effects in a multiplicative structure:

Separation Model: *There exist known volume measures $v_0, v_1, \ldots, v_n$ of the accident years as well as parameters $\lambda_0, \lambda_1, \ldots, \lambda_n, \lambda_{n+1}, \ldots, \lambda_{2n}$ and $\vartheta_0, \vartheta_1, \ldots, \vartheta_n$ with $\sum_{l=0}^{n} \vartheta_l = 1$ such that the identity*

$$E[Z_{i,k}] = v_i \, \lambda_{i+k} \, \vartheta_k$$

holds for all $i, k \in \{0, 1, \ldots, n\}$.

In the separation model

- v_i can be interpreted as an *accident year effect* of accident year i,
- λ_{i+k} as a *calendar year effect* of calendar year $i + k$ and
- ϑ_k as a *development year effect* of development year k.

Normally the calendar year effects originate at least partially from inflation. Therefore, volume measures influenced by inflation like premiums or sums insured are inappropriate for the analysis of the calendar year effects and a non-inflationary volume measure like the number of contracts or the expected number of claims should be chosen instead.

In this article we assume that the assumptions of the separation model are fulfilled and we interpret the volume measure v_i as the number of contracts in accident year i.

The separation method consists of three steps:

- Estimation of the parameters $\lambda_0, \lambda_1, \ldots, \lambda_n$ and $\vartheta_0, \vartheta_1, \ldots, \vartheta_n$ by the marginal sum method.
- Estimation of the parameters $\lambda_{n+1}, \ldots, \lambda_{2n}$ by extrapolation.
- Prediction of the future incremental losses.

As usual, predictors of reserves are then obtained by summing up the predictors of the corresponding incremental losses.

Normalization and Reflection of the Incremental Losses

To estimate the parameters $\lambda_0, \lambda_1, \ldots, \lambda_n$ and $\vartheta_0, \vartheta_1, \ldots, \vartheta_n$, the incremental losses are first normalized by means of the volume measures and then the *normalized incremental losses*

$$X_{i,k} := Z_{i,k}/v_i$$

are relabelled to obtain the *reflected normalized incremental losses*

$$Y_{i,k} := X_{n-i-k,k}$$

with $i, k \in \{0, 1, \ldots, n\}$ and $i + k \le n$. This relabelling corresponds to an inter-change between accident years and calendar years such that, for example, the last observable calendar year takes the place of the first accident year while the first observable calendar year takes the place of the last accident year.

Marginal Sum Method

The normalized incremental losses satisfy

$$E[X_{i,k}] = \lambda_{i+k}\,\vartheta_k$$

and it follows that the reflected normalized incremental losses satisfy

$$E[Y_{i,k}] = E[X_{n-i-k,k}] = \lambda_{n-i-k+k}\,\vartheta_k = \lambda_{n-i}\,\vartheta_k$$

Letting

$$\alpha_i := \lambda_{n-i}$$

yields

$$E[Y_{i,k}] = \alpha_i\,\vartheta_k$$

so that a *multiplicative model* exists for the reflected normalized incremental losses. In particular, the parameters $\vartheta_0, \vartheta_1, \ldots, \vartheta_n$ form a *development pattern for incremental quotas*.

The parameters of the multiplicative model can be estimated by the *marginal sum method*. The marginal sum estimators satisfy

$$\alpha_i^{MS} = S_{i,n-i}\prod_{l=n-i+1}^{n}\varphi_l^{CL}$$

and

$$\vartheta_k^{MS} = \begin{cases} \displaystyle\prod_{l=1}^{n} \frac{1}{\varphi_l^{CL}} & \text{if } k = 0 \\[3ex] \displaystyle\prod_{l=k+1}^{n} \frac{1}{\varphi_l^{CL}} - \prod_{l=k}^{n} \frac{1}{\varphi_l^{CL}} & \text{else} \end{cases}$$

with the *cumulative reflected normalized incremental losses*

$$S_{i,k} := \sum_{l=0}^{k} Y_{i,l}$$

and the corresponding *chain ladder factors*

$$\varphi_k^{CL} := \frac{\sum_{j=0}^{n-k} S_{j,k}}{\sum_{j=0}^{n-k} S_{j,k-1}}$$

This means that the marginal sum estimators in turn can be determined by the chain ladder method, applied to the cumulative reflected normalized incremental losses $S_{i,k}$, which yields $\alpha_i^{MS} = S_{i,n}^{CL}$.

For $c \in \{0, 1, \ldots, n\}$ define

$$\lambda_c^{MS} := \alpha_{n-c}^{MS} = S_{n-c,n}^{CL}$$

Then

$$\lambda_0^{MS}, \lambda_1^{MS}, \ldots, \lambda_n^{MS}$$

are estimators of the calendar year effects $\lambda_0, \lambda_1, \ldots, \lambda_n$ and

$$\vartheta_0^{MS}, \vartheta_1^{MS}, \ldots, \vartheta_n^{MS}$$

are estimators of the development year effects $\vartheta_0, \vartheta_1, \ldots, \vartheta_n$.

Example A. Consider the run-off triangle of incremental losses with the volume measures of the accident years:

Accident year i	Development year k						Volume v_i
	0	1	2	3	4	5	
0	1001	854	568	565	347	148	4025
1	1113	990	671	648	422		4456
2	1265	1168	800	744			5315
3	1490	1383	1007				5986
4	1725	1536					6939
5	1889						8158

We first pass to the run-off triangle of normalized incremental losses

Accident	Development year k					
year i	0	1	2	3	4	5
0	0.249	0.212	0.141	0.140	0.086	0.037
1	0.250	0.222	0.151	0.145	0.095	
2	0.238	0.220	0.151	0.140		
3	0.249	0.231	0.168			
4	0.249	0.221				
5	0.232					

and then to the run-off triangle of reflected normalized incremental losses:

Accident	Development year k					
year i	0	1	2	3	4	5
0	0.232	0.221	0.168	0.140	0.095	0.037
1	0.249	0.231	0.151	0.145	0.086	
2	0.249	0.220	0.151	0.140		
3	0.238	0.222	0.141			
4	0.250	0.212				
5	0.249					

Now the chain ladder method is applied to the cumulative reflected normalized incremental losses:

Accident	Development year k					
year i	0	1	2	3	4	5
0	0.232	0.453	0.621	0.761	0.856	0.893
1	0.249	0.480	0.630	0.776	0.862	*0.899*
2	0.249	0.469	0.619	0.760		*0.886*
3	0.238	0.460	0.601			*0.861*
4	0.250	0.462				*0.878*
5	0.249					*0.902*
φ_k^{CL}		*1.909*	*1.328*	*1.228*	*1.118*	*1.043*
γ_k^{CL}	*0.276*	*0.526*	*0.699*	*0.858*	*0.959*	*1*
ϑ_k^{CL}	*0.276*	*0.251*	*0.173*	*0.159*	*0.101*	*0.041*

Since $\vartheta_k^{MS} = \vartheta_k^{CL}$ holds for all $k \in \{0, 1, \ldots, 5\}$, the marginal sum estimators of the development year effects (and hence of the development pattern for incremental quotas) can be directly read off from the last row of the table; since $\lambda_c^{MS} = \alpha_{5-c}^{MS} = S_{5-c}^{CL}$ holds for all $c \in \{0, 1, \ldots, 5\}$, the marginal sum estimators of the calendar year effects are (in reverse order) contained in the last column and are given by the table

Calendar year c	0	1	2	3	4	5
λ_c^{MS}	0.902	0.878	0.861	0.886	0.889	0.893

Example B. In this example the incremental loss $Z_{4,1}$ is increased by 1000:

Accident year i	Development year k						Volume v_i
	0	1	2	3	4	5	
0	1001	854	568	565	347	148	4025
1	1113	990	671	648	422		4456
2	1265	1168	800	744			5315
3	1490	1383	1007				5986
4	1725	2536					6939
5	1889						8158

Run-off triangle of the normalized incremental losses:

Accident year i	Development year k					
	0	1	2	3	4	5
0	0.249	0.212	0.141	0.140	0.086	0.037
1	0.250	0.222	0.151	0.145	0.095	
2	0.238	0.220	0.151	0.140		
3	0.249	0.231	0.168			
4	0.249	0.365				
5	0.232					

Run-off triangle of the reflected normalized incremental losses:

Accident year i	Development year k					
	0	1	2	3	4	5
0	0.232	0.365	0.168	0.140	0.095	0.037
1	0.249	0.231	0.151	0.145	0.086	
2	0.249	0.220	0.151	0.140		
3	0.238	0.222	0.141			
4	0.250	0.212				
5	0.249					

Application of the chain ladder method to the cumulative reflected normalized incremental losses:

Accident	Development year k					
year i	0	1	2	3	4	5
0	0.232	0.597	0.765	0.905	1.000	1.037
1	0.249	0.480	0.630	0.776	0.862	0.893
2	0.249	0.469	0.619	0.760		0.872
3	0.238	0.460	0.601			0.836
4	0.250	0.462				0.838
5	0.249					0.915
φ_k^{CL}		2.028	1.304	1.211	1.108	1.037
γ_k^{CL}	0.272	0.551	0.719	0.871	0.965	1
$\vartheta_k^{\mathrm{CL}}$	0.272	0.279	0.168	0.152	0.094	0.035

The marginal sum estimators of the calendar year effects of the observable calendar years are given by the table

Calendar year c	0	1	2	3	4	5
λ_c^{MS}	0.915	0.838	0.836	0.872	0.893	1.037

The outlier $Z_{4,1}$ affects all estimators and the outlier effect is particularly significant in the estimator of the calendar year effect of calendar year 5.

The last example shows that data cleansing should also precede the application of the separation method.

Extrapolation

To estimate the calendar year effects of the non-observable calendar years, we need a model for the collection of all calendar year effects.

For example, the assumption of a geometric growth of the calendar year effects postulates the existence of a parameter δ such that the identity

$$\lambda_c = \delta \lambda_{c-1}$$

holds for all $c \in \{1, \ldots, 2n\}$. Minimization of the *squared approximation error*

$$\sum_{c=1}^{n} \left(\lambda_c^{\mathrm{MS}} - \delta \lambda_{c-1}^{\mathrm{MS}}\right)^2 = \sum_{c=1}^{n} (\lambda_{c-1}^{\mathrm{MS}})^2 \left(\frac{\lambda_c^{\mathrm{MS}}}{\lambda_{c-1}^{\mathrm{MS}}} - \delta\right)^2$$

results in the unique solution

$$\widehat{\delta} := \frac{\sum_{c=1}^{n} \lambda_{c-1}^{MS} \lambda_{c}^{MS}}{\sum_{c-1}^{n} \left(\lambda_{c-1}^{MS}\right)^2} = \sum_{c=1}^{n} \frac{\left(\lambda_{c-1}^{MS}\right)^2}{\sum_{d=1}^{n} \left(\lambda_{d-1}^{MS}\right)^2} \frac{\lambda_{c}^{MS}}{\lambda_{c-1}^{MS}}$$

Putting

$$\widehat{\lambda}_c := \widehat{\delta}^{c-n} \lambda_n^{MS}$$

then yields estimators of the calendar year effects λ_c with $c \in \{n+1, \ldots, 2n\}$.

Example A. From the table

Calendar year c	0	1	2	3	4	5
λ_c^{MS}	0.902	0.878	0.861	0.886	0.889	0.893

we obtain $\widehat{\delta} = 0.998$ and hence

Calendar year c	6	7	8	9	10
$\widehat{\lambda}_c$	0.890	0.888	0.886	0.884	0.882

In this case we get a *deflation* over all calender years.

Example B. From the table

Calendar year c	0	1	2	3	4	5
λ_c^{MS}	0.915	0.838	0.836	0.872	0.893	1.037

we obtain $\widehat{\delta} = 1.028$ and hence

Calendar year c	6	7	8	9	10
$\widehat{\lambda}_c$	1.065	1.095	1.125	1.156	1.188

In this case we get an *inflation* over all calender years.

These examples show that inflation or deflation effects can be completely reversed by outliers.

Quite generally, estimation of the calendar year effects of the non-observable calendar years can be carried out like the estimation of the *tail*.

Predictors of the Non-observable Incremental Losses

Because of the separation model, the predictors

$$Z_{i,k}^{SP} := v_i \,\widehat{\lambda}_{i+k}\, \vartheta_k^{MS}$$

are suitable to predict the future incremental losses $Z_{i,k}$ with $i, k \in \{1, \dots, n\}$ and $i + k \geq n + 1$.

Example A. We obtain the following table for the observable incremental losses and the predictors of the non-observable incremental losses:

Accident year i	Development year k					
	0	1	2	3	4	5
0	1001	854	568	565	347	148
1	1113	990	671	648	422	*163*
2	1265	1168	800	744	*478*	*194*
3	1490	1383	1007	*848*	*537*	*218*
4	1725	1536	*1066*	*980*	*621*	*253*
5	1889	*1820*	*1250*	*1150*	*728*	*296*

Example B. We obtain the following table for the observable incremental losses and the predictors of the non-observable incremental losses:

Accident year i	Development year k					
	0	1	2	3	4	5
0	1001	854	568	565	347	148
1	1113	990	671	648	422	*168*
2	1265	1168	800	744	*531*	*206*
3	1490	1383	1007	*969*	*614*	*239*
4	1725	2536	*1240*	*1154*	*732*	*284*
5	1889	*2428*	*1498*	*1394*	*884*	*344*

Again, the effect of the outlier is evident.

Notes

Keywords: Chain Ladder Method (Basics), Development Patterns (Basics), Marginal Sum Method, Multiplicative Models, Run-Off Data, Run-Off Triangles, Tail Estimation.

References: GDV [2011], Hossack, Pollard & Zehnwirth [1999], Institute of Actuaries [1989], Schmidt [2007], Taylor [1986, 2000], Van Eeghen [1981].

Simulation

Magda Schiegl

In the present article, we consider the use of the *Monte Carlo method* for the simulation of run-off triangles or run-off squares.

A given *run-off triangle* is a realization of a stochastic process, which describes the occurrence and the development of losses. If we assume that a stochastic process, and in particular the type and the parameters of its distribution, is given, then we can generate *run-off data*, i.e. realizations of the stochastic process, by means of the Monte Carlo method.

The assumption of a complete and realistic mathematical model describing the generation of run-off data is indispensable for the use of the Monte Carlo method.

The model considered in this article will be introduced in the following section. Of course, the Monte Carlo method can also be applied to other models. The high flexibility of the Monte Carlo method is per se its major advantage. In the application of the Monte Carlo method, the mathematical model is implemented as a computer programme and is in this way virtually brought to life.

The selected mathematical model represents a basic assumption to be aware of. However, once the model has been selected, then Monte Carlo simulation provides a significant advantage over the classical methods of loss reserving: Different to these, Monte Carlo simulation does not limit the evaluation of a risk to the calculation of a predictor and its prediction error but also allows for the determination of the complete distribution of a reserve; in particular, its quantiles can be determined. As a consequence, calculation of the capital at risk by the *value-at-risk* method is possible, which corresponds to the required Solvency II risk assessment.

If a mathematical model that describes the occurrence and the development of losses is assumed as given, then Monte Carlo simulation can be used to verify the results of classical methods of loss reserving. To this end, realizations of predictors

M. Schiegl (✉)
Hochschule Landshut, Landshut, Germany
e-mail: magda.schiegl@haw-landshut.de

© Springer International Publishing Switzerland 2016
M. Radtke et al. (eds.), *Handbook on Loss Reserving*,
EAA Series, DOI 10.1007/978-3-319-30056-6_35

of reserves for a run-off triangle generated by the Monte Carlo method are compared with the results of the Monte Carlo method for these reserves. It is then possible to make statistical statements on the quality of predictors of reserves for model-based Monte Carlo simulated data. In the section on backtesting we describe how properties of predictors of reserves (like unbiasedness) can be checked.

In the following we illustrate how the Monte Carlo method works, using a particular model for the occurrence and the development of losses.

The Model

The model for the generation of a run-off square presented here serves as a basis for the stochastic process that is realized by means of the Monte Carlo method. Of course, this model is just an example, which can be modified in various ways; see Schiegl (2009, 2015).

Consider the run-off square of incremental losses:

Accident year	Development year					
	0	1	... k	... $n-i$	... $n-1$	n
0	$Z_{0,0}$	$Z_{0,1}$	... $Z_{0,k}$	... $Z_{0,n-i}$	... $Z_{0,n-1}$	$Z_{0,n}$
1	$Z_{1,0}$	$Z_{1,1}$	... $Z_{1,k}$	... $Z_{1,n-i}$	... $Z_{1,n-1}$	$Z_{1,n}$
$\vdots$	$\vdots$	$\vdots$	$\vdots$	$\vdots$	$\vdots$	$\vdots$
i	$Z_{i,0}$	$Z_{i,1}$	... $Z_{i,k}$	... $Z_{i,n-i}$	... $Z_{i,n-1}$	$Z_{i,n}$
$\vdots$	$\vdots$	$\vdots$	$\vdots$	$\vdots$	$\vdots$	$\vdots$
$n-k$	$Z_{n-k,0}$	$Z_{n-k,1}$	... $Z_{n-k,k}$	... $Z_{n-k,n-i}$	... $Z_{n-k,n-1}$	$Z_{n-k,n}$
$\vdots$	$\vdots$	$\vdots$	$\vdots$	$\vdots$	$\vdots$	$\vdots$
$n-1$	$Z_{n-1,0}$	$Z_{n-1,1}$	... $Z_{n-1,k}$	... $Z_{n-1,n-i}$	... $Z_{n-1,n-1}$	$Z_{n-1,n}$
n	$Z_{n,0}$	$Z_{n,1}$	... $Z_{n,k}$	... $Z_{n,n-i}$	... $Z_{n,n-1}$	$Z_{n,n}$

We assume that the incremental losses $Z_{i,k}$ are observable for $i + k \leq n$ and that they are non-observable for $i + k \geq n + 1$. We consider the collective run-off model for the collection of all incremental losses:

Collective Run-Off Model:

(i) *For all $i, k \in \{0, 1, \ldots, n\}$, there exists a random variable $N_{i,k}$ with values in $\mathbb{N}_0$ and an independent and identically distributed sequence $\{X_{i,k,r}\}_{r \in \mathbb{N}}$ of random variables which is independent of $N_{i,k}$ and satisfies*

$$Z_{i,k} = \sum_{r=1}^{N_{i,k}} X_{i,k,r}$$

(ii) *The sequences $\{X_{i,k,r}\}_{r \in \mathbb{N}}$ with $i, k \in \{0, 1, \ldots, n\}$ are independent of each other.*

In this article we assume

- that the assumptions of the collective run-off model are fulfilled and
- that the distributions of the *claim sizes* $X_{i,k,r}$ are known.

Frequently used distributions for claim sizes are the *Gamma distribution* and the *Pareto distribution*.

The *incremental claim numbers* $N_{i,k}$ can be represented by a run-off square:

Accident year	Development year						
	0	1	... k	... n−i	... n−1	n
0	$N_{0,0}$	$N_{0,1}$	... $N_{0,k}$	... $N_{0,n-i}$	... $N_{0,n-1}$	$N_{0,n}$
1	$N_{1,0}$	$N_{1,1}$	... $N_{1,k}$	... $N_{1,n-i}$	... $N_{1,n-1}$	$N_{1,n}$
⋮	⋮	⋮	⋮	⋮	⋮	⋮
i	$N_{i,0}$	$N_{i,1}$	... $N_{i,k}$	... $N_{i,n-i}$	... $N_{i,n-1}$	$N_{i,n}$
⋮	⋮	⋮	⋮	⋮	⋮	⋮
n−k	$N_{n-k,0}$	$N_{n-k,1}$	... $N_{n-k,k}$	... $N_{n-k,n-i}$	... $N_{n-k,n-1}$	$N_{n-k,n}$
⋮	⋮	⋮	⋮	⋮	⋮	⋮
n−1	$N_{n-1,0}$	$N_{n-1,1}$	... $N_{n-1,k}$	... $N_{n-1,n-i}$	... $N_{n-1,n-1}$	$N_{n-1,n}$
n	$N_{n,0}$	$N_{n,1}$	... $N_{n,k}$	... $N_{n,n-i}$	... $N_{n,n-1}$	$N_{n,n}$

For $i \in \{0, 1, \ldots, n\}$, define

$$N_i := \sum_{k=0}^{n} N_{i,k}$$

and

$$\alpha_i := E[N_i]$$

Then the *ultimate claim number* N_i is the number of all claims from accident year i. For the collection of all claim numbers we consider the multinomial model with independent accident years:

Multinomial Model with Independent Accident Years:

  (i) *The random variables $N_{i,k}$ take only values in $\mathbb{N}_0$.*

 (ii) *The families $\{N_{i,k}\}_{k \in \{0,1,\ldots,n\}}$ are independent of each other.*

(iii) *There exist parameters $\vartheta_0, \vartheta_1, \ldots, \vartheta_n \in (0, 1)$ with $\sum_{l=0}^{n} \vartheta_l = 1$ such that*

$$P_{N_{i,0}, N_{i,1}, \ldots, N_{i,n} | N_i} = \mathbf{M}(N_i ; \vartheta_0, \vartheta_1, \ldots, \vartheta_n)$$

holds for all $i \in \{0, 1, \ldots, n\}$.

In this article we also assume

- that the assumptions of the multinomial model with independent accident years are fulfilled,
- that the *development pattern for incremental quotas* $\vartheta_0, \vartheta_1, \ldots, \vartheta_n$ is known and
- that the distributions of the ultimate claim numbers are known.

Frequently used distributions for claim numbers are the *binomial distribution*, the *Poisson distribution* and the *negative binomial distribution* (and thus the distributions of the *Panjer class*).

The model considered here contains the *multiplicative model for incremental losses* as a special case: If we assume in addition that there exist parameters $\mu_0, \mu_1, \ldots, \mu_n$ such that the identity

$$E[X_{i,k,r}] = \mu_i$$

holds for all $k \in \{0, 1, \ldots, n\}$ and $r \in \mathbb{N}$, then the identity

$$E[Z_{i,k}] = E[N_{i,k}] \, E[X_{i,k,r}] = \vartheta_k \, E[N_i] \, \mu_i = \vartheta_k \, \alpha_i \, \mu_i = (\alpha_i \mu_i) \, \vartheta_k$$

holds for all $i, k \in \{0, 1, \ldots, n\}$. In particular, we obtain the *Poisson model* if we assume that all ultimate claim numbers have a Poisson distribution and that all claim sizes are equal to 1.

The Monte Carlo Method

The Monte Carlo method enables the simulation of an arbitrary stochastic process. For this purpose a real process (in the present case the occurrence and the development of losses) is reduced by a mathematical model to its essential properties and relationships. This model is implemented as a computer code by means of the Monte Carlo method and is in this way virtually brought to life. In the computerized model relationships and dependencies can be studied in order to gain insights into the real world.

We describe this approach using the example of our model for the generation of a run-off square:

Using the Monte Carlo method we generate realizations of the claim numbers N_i and $N_{i,k}$ and of the claim sizes $X_{i,k,r}$. These realizations correspond to *random numbers* underlying the distributions of the random variables. For practical reasons, however, truly random numbers are generally not used in practice and are replaced by *pseudorandom numbers*, which are generated by appropriate algorithms.

Standard software provides the generation of pseudorandom numbers on the unit interval. Pseudorandom numbers must satisfy certain conditions in order to imitate truly random numbers as well as possible. For example, pseudorandom numbers have to be uniformly distributed on the unit interval and have to be independent, and even over long periods systematic repetitions of a sequence of pseudorandom numbers or other dependencies must not occur.

The real art of the Monte Carlo simulation consists in the transformation of pseudorandom numbers on the unit interval into pseudorandom numbers that obey a given distribution. This transformation is briefly outlined here:

Let F be a distribution function which is strictly monotonically increasing on the set $J_F := \{x \in \mathbb{R} \mid 0 < F(x) < 1\}$ and let U be a random variable having the *uniform distribution* on the unit interval. Then the random variable

$$X := F^{-1}(U)$$

has the distribution function F since

$$P[\{X \leq x\}] = P[\{F^{-1}(U) \leq x\}] = P[\{U \leq F(x)\}] = F(x)$$

This principle for the generation of pseudorandom numbers with a given distribution function F can be also applied, with a slightly changed definition of the random variable X, if the distribution function on J_F is not strictly monotonically increasing.

Example. The distribution function of the *Pareto distribution* **Par**(ξ, β) is strictly monotonically increasing on the interval $J_F = (\xi, \infty)$ since

$$F(x) = \begin{cases} 0 & \text{if } x \leq \xi \\ 1 - \left(\dfrac{\xi}{x}\right)^{\beta} & \text{if } x > \xi \end{cases}$$

This yields

$$F^{-1}(u) = \frac{\xi}{\sqrt[\beta]{1-u}}$$

for all $u \in (0, 1)$.

In many cases, however, the function F^{-1} cannot be given in closed form and its numerical computation by standard methods may be quite time-consuming, but often a faster method can be found in the literature.

Kalos and Whitelock (2008) give a useful summary of the generation and the testing of pseudorandom numbers as well as of the Monte Carlo method. For actuarial applications we also recommend Daykin et al. (1994).

Implementation by a Computer Code

The model is implemented by a computer code in order to obtain realizations of the
stochastic process which describes the occurrence and the development of losses.
The computer code follows an algorithm:

(1) For every accident year $i \in \{0, 1, \ldots, n\}$, a realization n_i of the ultimate claim
    number N_i is generated.
(2) Using a development pattern $\vartheta_0, \vartheta_1, \ldots, \vartheta_n$ for the incremental claim numbers,
    the multinomial distribution $\mathbf{M}(n_i; \vartheta_0, \vartheta_1, \ldots, \vartheta_n)$ is used to produce realiza-
    tions $n_{i,k}$ of the incremental claim numbers $N_{i,k}$ with $k \in \{0, 1, \ldots, n\}$ such
    that

$$n_i := \sum_{k=0}^{n} n_{i,k}$$

(3) Realizations $x_{i,k,1}, \ldots, x_{i,k,n_{i,k}}$ of the claim sizes $X_{i,k,1}, \ldots, X_{i,k,n_{i,k}}$ are gener-
    ated for every accident year i and development year k and a realization of the
    incremental loss $Z_{i,k}$ is then obtained by letting

$$z_{i,k} := \sum_{r=1}^{n_{i,k}} x_{i,k,r}$$

Once this has been done, a realization of the run-off square for incremental losses is
available and can be used to obtain realizations of related random variables:

(4) A realization of the cumulative loss $S_{i,k} := \sum_{l=0}^{k} Z_{i,l}$ is obtained by letting

$$s_{i,k} := \sum_{l=0}^{k} z_{i,l}$$

(5) A realization of the accident year reserve $R_i := S_{i,n} - S_{i,n-i}$ for accident year i
    is obtained by letting

$$r_i := s_{i,n} - s_{i,n-i}$$

In the same way we obtain realizations of the calendar year reserves and of the
aggregate loss reserve.

Statistical Statements

A large number of realizations is needed to make statistical statements on the distribution of the random variable under consideration. This means that the single steps of the computer algorithm have to be repeated very often (depending on the task and on the required precision of the result about 10 000 times or even more). From the whole set of Monte Carlo samples we obtain an empirical distribution, which is called the *Monte Carlo distribution* in the following.

The Monte Carlo distribution of a random variable yields, for instance, approximations of its first and second moments. This, however, is of minor importance in the model considered here, since the first and second moments of the incremental and cumulative losses and those of the reserves can be determined explicitly by means of the first and second moments of the ultimate claim number and of the claim sizes.

The substantial power of the Monte Carlo method consists in the possibility to approximate not only certain moments, but even the whole distribution of a random variable and hence, in particular, its quantiles. For example, we can answer the question of how much has to be reserved in addition to the expectation in order to avoid being under-reserved with a given probability like 85 % or 99.5 %. These two quantiles are important for the calculation of the *minimum capital requirement* (MCR) and the *solvency capital requirement* (SCR), respectively, in the Solvency II framework.

With the quantiles of the reserve distribution it is possible to quantify the risk that the reserve is not sufficient to cover the liabilities. Thus the technical provisions that are relevant for the Solvency II valuation are accessible by the risk measure *value-at-risk*. With the described method other risk measures like the *tail value-at-risk* can also be calculated for internal risk assessment approaches.

Quality Control of the Results

In order to ensure that the numerical results of the computer code are consistent with the model we have to perform some tests. In doing so, we assume that the random numbers provided by the computer are appropriate for the calculations to be carried out; this cannot be taken for granted at all, as pointed out before, and hence also has to be checked in practice.

First of all we make sure that the random numbers generated by the Monte Carlo algorithms obey the given distributions. We then check the correct split of the ultimate loss number into the incremental loss numbers according to the multinomial distribution and the correct calculation of the distributions of the incremental losses. Furthermore, it should be checked whether the results stabilize with an increasing number of Monte Carlo samples and hence satisfy the law of large numbers.

As a representative of these testing procedures we describe here one testing procedure in detail and present the results: We consider an incremental loss

$$Z_{i,k} = \sum_{r=1}^{N_{i,k}} X_{i,k,r}$$

and use a Monte Carlo algorithm to simulate the distribution of $Z_{i,k}$. As the incremental loss is the aggregate loss in a collective model, we can compare the Monte Carlo distribution with the distribution provided by *Panjer's recursion* in order to check the quality of the simulation.

Example. We specify the distributions of the claim numbers and of the claim sizes as follows:

- We choose a development pattern for incremental quotas $\vartheta_0, \vartheta_1, \ldots, \vartheta_n$.
- We choose the *Poisson distribution* $\mathbf{Poi}(\alpha_i)$ for the ultimate claim number N_i of accident year i. Then, for all $i, k \in \{0, 1, \ldots, n\}$, the incremental claim number $N_{i,k}$ has the Poisson distribution $\mathbf{Poi}(\alpha_i \vartheta_k)$.
- We choose the *Pareto distribution* $\mathbf{Par}(\xi_{i,k}, \beta_{i,k})$ for the claim sizes $X_{i,k,r}$ from accident year i and in development year k.

As the Pareto distribution is a continuous distribution, we discretize the Pareto distribution with step width h by letting $f_0 := 0$ and

$$f_n := P\left[\left\{(n-1)\,h < X_{i,k,r} \leq n\,h\right\}\right]$$

for all $n \in \mathbb{N}$. Then Panjer's recursion, given by $g_0 := e^{-\alpha_i \vartheta_k}$ and

$$g_n := (\alpha_i \vartheta_k) \sum_{j=1}^{n} \frac{j}{n}\, g_{n-j}\, f_j$$

for all $n \in \mathbb{N}$, produces a discrete distribution which approximates the distribution of the incremental loss $Z_{i,k}$. This approximation of the distribution of $Z_{i,k}$ can be compared with its Monte Carlo distribution.

The following figure shows the quantiles (black squares) and the standard deviations of the quantiles (white squares) of the Monte Carlo distribution for $\alpha_i = 120$ and $\vartheta_k = 0.1$ as well as $\xi_{i,k} = 1000$ and $\beta_{i,k} = 1.1$ on the basis of 1000 Monte Carlo realizations and in addition the approximated distribution of the incremental loss $Z_{i,k}$ generated by Panjer's recursion with step width $h = 50$ (line):

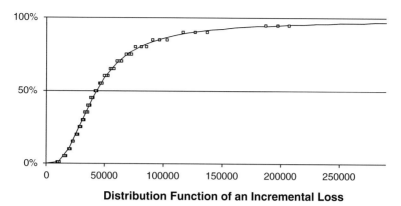

Distribution Function of an Incremental Loss

We have $E[N_{i,k}] = 12 = \text{var}[N_{i,k}]$ as well as $E[X_{i,k,r}] = 11\,000$ and $\text{var}[X_{i,k,r}] = \infty$, and hence $E[Z_{i,k}] = 132\,000$ and $\text{var}[Z_{i,k}] = \infty$.

Estimation of Reserves

The previous considerations start with a stochastic model that generates a run-off triangle of incremental losses.

By contrast, for a given (numerical) run-off triangle, the question arises which stochastic model provides an appropriate description of how the data of the run-off triangle are generated. The choice of such a stochastic model should take into account not only the (internal) information contained in the run-off triangle, but also the whole data set of all individual losses leading to the run-off triangle as well as any (external) information about the given segment that can be gained from similar segments or market statistics. Once the type of a parametric model has been selected, calibration using the data provides plausible values for the parameters of the model.

For a given run-off triangle for which, after careful review using all available information, it can be assumed that it is described appropriately by a certain stochastic model, the Monte Carlo method can be used to estimate reserves. This is done by steps (4) and (5) of the algorithm described above.

The advantage of reserve estimation by means of the model-based Monte Carlo method consists in the fact that it generates simultaneously and in a consistent manner the distribution of the aggregate loss reserve and those of all partial reserves (like accident year and calendar year reserves), and hence their quantiles, whereas classical methods of loss reserving provide only predictors and prediction errors.

Backtesting of a Loss Reserving Method

Under the assumption of a specific stochastic model the Monte Carlo method can also be used to study the properties of a loss reserving method.

For backtesting a loss reserving method and a reserve estimator provided by this method, we compute

- Monte Carlo realizations of the reserve estimator by applying the loss reserving method to the Monte Carlo realizations of the (upper) run-off triangle and
- Monte Carlo realizations of the reserve by using the Monte Carlo realizations of the lower triangle of the run-off square.

Comparing the statistical properties of the reserve estimator with those of the Monte Carlo reserve we can get some insight into the quality of the reserve estimator under the assumptions of the model. For example, conclusions about unbiasedness of the reserve estimator or the reliability of a confidence interval for the reserve estimator can be drawn in this way.

Backtesting is most important in practice since certain properties of reserve estimators (and also those of estimators of their estimation errors) can only be proven under certain model assumptions that need not be fulfilled for a given portfolio of risks. For example, unbiasedness of the *chain ladder predictors* has been proven only under the assumptions of the *chain ladder model of Mack* or those of the *chain ladder model of Schnaus*, and their optimality for the first non-observable calendar year has been proven only under the assumptions of the *sequential conditional linear chain ladder model*.

The Monte Carlo method and backtesting was already applied to study the properties of the chain ladder method; see Schiegl (2002, 2015).

Notes

Keywords: Chain Ladder Method (Basics), Chain Ladder Method (Models), Collective Model, Development Patterns (Basics), Multinomial Model, Multiplicative Model, Poisson Model, Run-Off Data, Run-Off Triangles.

Literature: Daykin, Pentikäinen & Pesonen [1994], Kalos & Whitelock [2008], Mack [2002], Schiegl [2002, 2009, 2015], Schmidt & Schnaus [1996].

Solvency II

Alexander Ludwig

Since 1 January 2016 Solvency II is the binding regulatory law for insurers and reinsurers in the European Economic Area.[1] Solvency II captures requirements on the calculation of regulatory risk capital and own funds, on the system of governance and internal risk management as well as on disclosure to the supervisory authorities and the public.

This article presents the main principles of Solvency II and its standard formula, the widely used canonical approach to implement the requirements on the calculation of the *solvency capital requirement* (SCR). Moreover, this article highlights the necessity to integrate loss reserving methods when determining the Solvency II balance sheet and the derivation of the SCR. This is in line with the scope of this handbook on non-life insurance business.

Motivation and Elements of Solvency II

Solvency II demands a risk-based management of (re)insurance undertakings.[2] It is based on a three pillar model similar to the approach for banks under Basel II/III:

- The first pillar defines a methodological framework for the derivation of capital requirements for (re)insurers.

---

[1] All EU member states as well as Iceland, Liechtenstein and Norway.

[2] Directive on the taking-up and pursuit of the business of Insurance and Reinsurance (Solvency II), DIRECTIVE 2009/138/EC, Article 44 (Solvency II Directive). In the following, references to legal texts will always be given with respect to their first appearance within the Lamfalussy process to establish Solvency II.

A. Ludwig (✉)
Technische Universität Dresden, Dresden, Germany
e-mail: alexander.ludwig2@googlemail.com

© Springer International Publishing Switzerland 2016
M. Radtke et al. (eds.), *Handbook on Loss Reserving*,
EAA Series, DOI 10.1007/978-3-319-30056-6_36

- The second pillar contains requirements on the system of governance and internal risk management.
- The third pillar captures requirements on the disclosure of information regarding the solvency position of the (re)insurer to the supervisory authorities and the public.

In the following the three pillars will be presented briefly.

Pillar 1: Solvency Balance Sheet and Capital Requirements

The first pillar of Solvency II defines principles for the valuation of (re)insurer's assets, technical provisions and other liabilities as well as for the determination of

- the *solvency capital requirement* (SCR)[3] and
- the *minimum capital requirement* (MCR).[4]

The (re)insurance undertaking shall possess sufficient eligible *own funds* to cover the SCR,[5] and it is obliged to possess sufficient eligible own funds to cover the MCR.[6] In the following, the article highlights essential principles of Pillar 1.

Market-Consistent Valuation

Solvency II requires a market-consistent valuation of assets, technical provisions and other liabilities. Market-consistent values are either market values observable on liquid financial markets (*mark-to-market* valuation) or are estimated using appropriate valuation models (*mark-to-model* valuation).

The value of technical provisions shall reflect the price a (re)insurance undertaking would have to pay when immediately transferring their (re)insurance liabilities to another (re)insurance undertaking.[7] Technical provisions are valuated as the sum of a best estimate and a risk margin[8] and shall be calculated on the level of homogeneous risk groups.[9] The *best estimate* of the insurance liabilities is the expected present value of future gross cashflows discounted by risk-free interest rates.[10] The

---

[3]Solvency II Directive, Articles 100–127.
[4]Solvency II Directive, Articles 128–131.
[5]Solvency II Directive, Article 100.
[6]Solvency II Directive, Article 128.
[7]Solvency II Directive, Article 76 (2).
[8]Solvency II Directive, Article 77 (1).
[9]Solvency II Directive, Article 80.
[10]Solvency II Directive, Article 77 (2). Reinsurance recoverables, i.e. the cedent's expected present value of benefits from reinsurance contracts, are captured as an asset.

relevance of actuarial models is underlined by Solvency II as *adequate, applicable and relevant actuarial and statistical methods* have to be used for the determination of best estimates.[11]

The risk margin equals the value which another insurance undertaking (*reference undertaking*) would ask for—besides the best estimate—to take over the insurance liabilities and associated risks. If the (re)insurer calculates the risk margin separately from the best estimate, the risk margin shall equal the present value of expected future capital costs which would arise from covering the risks of the insurance liabilities until their final run-off.[12] Thus, to complete the Solvency II balance sheet, a projection of risks until the final run-off of the business in force has to be performed, among other things.

Determination of Eligible Own Funds

Based on the market-consistent valuation of assets, technical provisions and other liabilities in the Solvency II balance sheet, unrealized gains and losses in comparison to statutory accounting rules usually emerge. In the case of unrealized gains, the (re)insurance undertaking has to recognize *deferred tax liabilities* in the balance sheet, whereas in the case of unrealized losses, the undertaking can recognize *deferred tax assets* if it is able to demonstrate that deferred tax assets may be utilized against future tax liabilities arising from future profits.

The following table shows an exemplary aggregated Solvency II balance sheet (in million euro):

Assets	Value	Liabilities	Value
Real estate	2,990	Excess of assets over liabilities	21,103
Equity	8,675	Best estimates	156,297
Bonds	165,750	Risk margin	1,249
Reinsurance recoverables	4,200	Other liabilities	390
Deferred tax assets	1,952	Deferred tax liabilities	4,528
Total	183,567	Total	183,567

Own funds eligible to cover the SCR/MCR are derived from the *excess of assets over liabilities* in the Solvency II balance sheet taking into account quality constraints on own funds[13] and certain deductions such as own shares and foreseeable dividends and distributions.

---

[11] Solvency II Directive, Article 77 (2).

[12] Solvency II Directive, Article 77 (5). The cost-of-capital rate shall equal 6 %, Delegated Regulation (EU) 2015/35, Article 39.

[13] Solvency II Directive, Article 93–95.

Solvency Capital Requirement (SCR)

The SCR shall represent the 99.5 % quantile of the distribution of own funds of the (re)insurance undertaking with respect to a one-year horizon.[14] To align with the requirement of prospective risk measurement and the going-concern-principle[15] the risk of the business in force as well as of new business of the next 12 months shall be considered.[16] The derivation of the SCR can be performed using either the so-called *standard formula* or an *internal model*, which needs to be certified by the responsible supervisory authorities.[17]

The standard formula is used by the majority of (re)insurance undertakings and will be explained in more detail below.

Internal models can be partial internal or full internal models. The modular design of the standard formula allows us to switch from the standard formula to a company-individual modelling for different sets of risk (sub-)modules. Internal models are not within the scope of this article.

Minimum Capital Requirement (MCR)

The determination of the MCR is factor-based taking into account

- an absolute floor depending on the type of insurance business,[18]
- an MCR floor of 25 % of the SCR and an MCR cap of 45 % of the SCR[19] and
- a linear function of net best estimates and booked net premiums of the last 12 months (in case of non-life insurance liabilities) and net best estimates and capital-at-risk (in the case of life insurance liabilities).[20]

Pillar 2: System of Governance

All (re)insurance undertakings must have in place *an effective system of governance which provides for sound and prudent management of the business.*[21] This system

---

[14]Solvency II Directive, Article 101 (3). In other words, the (re)insurer has to determine the loss of own funds in a one-in-200-years event taking into account all quantifiable risks and their interrelations.

[15]Solvency II Directive, Article 101 (2).

[16]Solvency II Directive, Article 101 (3).

[17]Solvency II Directive, Article 100 and Article 112.

[18]Solvency II Directive, Article 129 (1).

[19]Solvency II Directive, Article 129 (3).

[20]Delegated Regulation (EU) 2015/35, Article 250.

[21]Solvency II Directive, Article 41(1).

shall include an effective internal control system, an objective internal audit function, an actuarial function, a compliance function as well as a risk management function. The management of the undertaking has to fulfill fit and proper requirements and the risk-management shall be integrated into the organizational structure and the decision-making process of the undertaking.[22]

Pillar 2 also requires the *own risk and solvency assessment* (ORSA), which especially demands an evaluation of whether the (re)insurance undertaking will have sufficient own funds to cover the SCR and the *own solvency needs* (OSN) on a continuous basis. The OSN reflects the essential undertaking-specific risks and takes into account undertaking-specific risk quantification methods. Moreover, for standard formula users the ORSA requires a critical assessment of the validity of the assumption of the standard formula with regard to the undertaking's risk profile.[23]

Pillar 3: Disclosure

The third pillar contains requirements on the disclosure of (re)insurance undertakings, which aim at an increase of transparency on the European insurance market and eventually in an increase of policyholder protection. Undertakings have to annually disclose the *solvency and financial condition report* (SFCR), which shall inform the public about its risk and capital management. Further, a qualitative report to the supervisory authority (*regular supervisory report*, (RSR) shall contain the statements of the SFCR but provide information in a degree of detail *which is necessary for the purposes of supervision.*[24] Both reports shall be accompanied with a prespecified list of *quantitative reporting templates* (QRT) for solo undertakings and insurance groups, which partially have to be disclosed on a quarterly frequency.[25]

Standard Formula

In the following, this article first explains the standard formula for the determination of the SCR. Afterwards the quantification of premium and reserve risk, which is the most relevant module with respect to actuarial reserving methods, within the standard formula, will be explained in detail.

---

[22]See Solvency II Directive, Articles 42–49, for all requirements as well as the EIOPA Guidelines on system of governance, EIOPA–BoS–14/253.

[23]See EIOPA Guidelines on own risk and solvency assessment, EIOPA–BoS–14/259, for further information.

[24]Solvency II Directive, Article 35 (1). See Delegated Regulation (EU) 2015/35, Annex XX, and EIOPA Guidelines for reporting and public disclosure, EIOPA–BoS–15/109, for details on SFCR and RSR.

[25]See EIOPA Implementing Technical Standards on the templates for the submission of information to the supervisory authorities, EIOPA–BoS–15/115, for details.

Determination of the SCR Using the Standard Formula

The following risk modules are considered separately in the standard formula[26]:

- market risk
- health underwriting risk
- counterparty default risk
- life underwriting risk
- non-life underwriting risk
- intangible asset risk
- operational risk

Most of the risk modules contain several sub-modules. The following figure displays the risk aggregation under Solvency II:

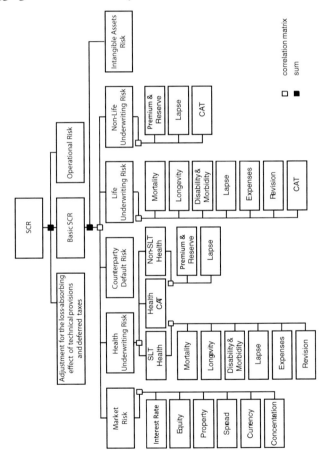

---

[26]Delegated Regulation (EU) 2015/35.

In the standard formula, capital requirements for each risk (sub-)module such as the premium and reserve risk are either determined by factor-based approaches or by shock scenarios:

- In factor-based approaches a volume measure, e.g. the best estimate of claims reserves of a line of business, is multiplied by a risk factor, e.g. the volatility of the claims development result on a 12-month horizon. Risk factors have been calibrated by EIOPA on the basis of market data and industry averages such that they shall represent a *value-at-risk* at a confidence level of 99.5 %.
- The application of shock scenarios yields the excess of assets over liabilities (net asset value) before and after a shock, and the difference between those values then provides the capital requirement for the respective risk (sub-)module. For example, property risk is quantified assuming a downturn of real estate values by 25 %.

Capital requirements on the level of risk sub-modules and risk modules, respectively, are aggregated using correlations and yield the *basic solvency capital requirement* (BSCR).

The final SCR results from the summation of the BSCR with capital requirements for operational risk, the loss absorbing capacity of technical provisions, i.e. the insurer's discretion to reduce part of policyholders' future benefits,[27] and the loss absorbing capacity of deferred taxes.[28] The latter is motivated by the fact that the undertaking might benefit from further deferred tax assets after the occurrence of the one-in-200-years loss event, which needs to be demonstrated by the undertaking.

Premium and Reserve Risk

Premium risk results from unexpected changes in the height, frequency and time of occurrence of claims and claims regulation costs with regard to future claim events. The reserve risk under Solvency II is the risk of a negative claims development result within the next 12 months, i.e. the unexpected need to recognize further provisions for claims which already occurred.

The capital requirement is determined by a joint factor-based approach for both premium and reserve risk.[29] To this end, an overall volume measure V and an overall volatility σ need to be calculated. The capital requirement results, using the formula

$$\mathrm{SCR}_{p/r} := 3\, V\sigma$$

which yields an approximation of the 99.5 % quantile of a lognormal distribution.

---

[27]Delegated Regulation (EU) 2015/35, Article 206.

[28]Delegated Regulation (EU) 2015/35, Article 207.

[29]See Delegated Regulation EU (2015/35), Articles 115–117, for the legal background of this section.

Determination of the Overall Volume Measure

The overall volume measure equals the sum of volume measures V_{lob} of each *line of business* (LOB)

$$V := \sum_{lob} V_{lob}$$

with

$$V_{lob} := \left(V_{(prem,lob)} + V_{(res,lob)} \right) \left(0.75 + 0.25\, DIV_{lob} \right)$$

i.e. the volume measure of a single line of business is determined from the volume measures $V_{(prem,lob)}$ for the premium risk and $V_{(res,lob)}$ for the reserve risk and the *Herfindahl index*

$$DIV_{lob} := \frac{\sum_j \left(V_{(prem,j,lob)} + V_{(res,j,lob)} \right)^2}{\left(\sum_j \left(V_{(prem,j,lob)} + V_{(res,j,lob)} \right) \right)^2}$$

The Herfindahl index enables the consideration of diversification effects between country regions such as Western Europe, Southern Europe, etc. The volume measure for the reserve risk of a line of business equals the net best estimate of claims reserves, and the volume measure for the premium risk of a line of business is the sum of

- the maximum of earned net premiums of the last 12 months and the expected earned net premiums of the next 12 months,
- the present value of expected net earned premiums for the period after the next 12 months for business in force, and
- the present value of expected net earned premiums for the period after the next 12 months for new business written in the next 12 months.

Determination of the Overall Volatility

The overall volatility

$$\sigma := \frac{1}{V} \left(\sum_{lob_1} \sum_{lob_2} Corr_{lob_1,lob_2} \cdot \sigma_{lob_1} \cdot \sigma_{lob_2} \cdot V_{lob_1} \cdot V_{lob_2} \right)^{1/2}$$

results from the volatilities of single lines of business, correlations between the lines of business and the respective volume measures. Correlations $Corr_{lob_1,lob_2}$ between lines of business are prespecified by the standard formula and the volatility of a line of business is given by

$$
\sigma_{lob} := \frac{1}{V_{(prem,lob)} + V_{(res,lob)}} \Big(\big(\sigma_{(prem,lob)} \cdot V_{(prem,lob)} \big)^2
$$
$$
+ 2 \cdot \alpha \cdot \sigma_{(prem,lob)} \cdot \sigma_{(res,lob)} \cdot V_{(prem,lob)} \cdot V_{(res,lob)}
$$
$$
+ \big(\sigma_{(res,lob)} \cdot V_{(res,lob)} \big)^2 \Big)^{1/2}
$$

Volatilities $\sigma_{(prem,lob)}$ and $\sigma_{(res,lob)}$ for premium risk and reserve risk, respectively, are also prespecified by the standard formula and shall reflect the differences in the risk proneness of the lines of business. The following table contains these volatilities for important lines of business[30]:

Line of Business	$\sigma_{(prem,lob)}$	$\sigma_{(res,lob)}$
Motor vehicle liability	10.0%	9.0%
Other motor	8.0%	8.0%
General liability	14.0%	11.0%

The correlation α between premium and reserve risk is set to $\alpha := 0.5$.

Undertaking-Specific Parameters (USP)

Premium risk and reserve risk volatilities can be replaced by *undertaking-specific parameters* (USP) within the standard formula. The use of undertaking-specific parameters has to be approved by the supervisory authorities. The following approaches can be used to determine undertaking-specific volatilities of a line of business: XVII.[31]

- premium risk method (lognormal)
- reserve risk method 1 (lognormal)
- reserve risk method 2 (Merz and Wüthrich)

If an undertaking is able to apply two different methods for reserve risk, it shall identify the method that better fits the calibration to a value-at-risk of 99.5%. If an undertaking is not able to identify this method, it shall use the more conservative approach.

Moreover, the volatility of premium risk can be reduced by a multiplication with an adjustment factor representing the risk-mitigating effects of non-proportional reinsurance contracts. Besides the quantification of this adjustment factor by a USP method,[32] XVII. undertakings can make use of a default setting of 80% for this

---

[30]Delegated Regulation (EU) 2015/35, Annex II.
[31]Delegated Regulation (EU) 2015/35, Article 220 and Annex XVII.
[32]Delegated Regulation (EU) 2015/35, Article 220 and Annex XVII.

adjustment factor in the lines of business motor vehicle liability, fire and other property damage and general liability, if their obligations in these lines of business are reinsured by an excess of loss reinsurance contract which fulfills the requirements of the Delegated Regulation (EU) 2015/35, Article 218 (2).

Notes

Reference: Fuchs, Ludwig & Schmidt [2013].

Tail Estimation

Heinz J. Klemmt, Michael Radtke and Anja Schnaus

The extension of the run-off square of cumulative losses by d additional development years yields the following *run-off rectangle*:

Accident	Development year							
year	0	1	... k	... n	$n+1$	... $n+d$		
0	$S_{0,0}$	$S_{0,1}$	... $S_{0,k}$	... $S_{0,n}$	$S_{0,n+1}$	... $S_{0,n+d}$		
1	$S_{1,0}$	$S_{1,1}$	... $S_{1,k}$	... $S_{1,n}$	$S_{1,n+1}$	... $S_{1,n+d}$		
⋮	⋮	⋮		⋮	⋮	⋮		
$n-k$	$S_{n-k,0}$	$S_{n-k,1}$	... $S_{n-k,k}$	... $S_{n-k,n}$	$S_{n-k,n+1}$	... $S_{n-k,n+d}$		
⋮	⋮	⋮	⋮	⋮	⋮	⋮		
$n-1$	$S_{n-1,0}$	$S_{n-1,1}$	... $S_{n-1,k}$	... $S_{n-1,n}$	$S_{n-1,n+1}$	... $S_{n-1,n+d}$		
n	$S_{n,0}$	$S_{n,1}$	... $S_{n,k}$	... $S_{n,n}$	$S_{n,n+1}$	... $S_{n,n+d}$		

We assume that the cumulative losses $S_{i,k}$ are observable for $i + k \leq n$ and that they are non-observable for $i + k \geq n + 1$ and that the development is not yet finished after n years but after d further years. The further development from development year $n + 1$ until development year $n + d$ is called the *tail*.

In practice the tail has to be considered if

- even the development of the eldest accident year cannot be observed completely (for example, in new portfolios or in long developing (*long-tail*) lines of business) or
- the observation period has to be restricted to a subsection because of systematic breaks in the *run-off data* (for example, in case of highly erroneous, particularly

H.J. Klemmt · M. Radtke · A. Schnaus (✉)
Fachhochschule Dortmund, Dortmund, Germany
e-mail: michael.radtke@towerswatson.com

© Springer International Publishing Switzerland 2016
M. Radtke et al. (eds.), *Handbook on Loss Reserving*,
EAA Series, DOI 10.1007/978-3-319-30056-6_37

older, data, in case of portfolio changes due to modified deductibles or liability limits, or in case of an entirely changed reinsurance structure).

In this article we consider the tail for methods of loss reserving that are based on development patterns. These include all methods that are subject to the Bornhuetter–Ferguson principle, especially the *Bornhuetter–Ferguson method*, the *loss development method*, the *chain ladder method*, the *Cape Cod method*, the *additive method* and the *Panning method*.

Extended Development Pattern for Factors

We extend the *development pattern for factors*:

> **Extended Development Pattern for Factors:** *There exist parameters* $\varphi_1, \ldots, \varphi_n,$ $\varphi_{n+1}, \ldots, \varphi_{n+d}$ *such that the identity*
>
> $$\frac{E[S_{i,k}]}{E[S_{i,k-1}]} = \varphi_k$$
>
> *holds for all* $k \in \{1, \ldots, n+d\}$ *and for all* $i \in \{0, 1, \ldots, n\}$.

In the same way, the development patterns for quotas and incremental quotas can be extended and it is, as in the case $d = 0$, easy to see that these three development patterns are equivalent. Also, in the same way as the extended development patterns, estimators of the parameters of one of these extended development patterns can be converted into estimators of the parameters of any of the other two extended development patterns.

In this article we assume that there exists an extended development pattern $\varphi_1, \ldots, \varphi_n, \varphi_{n+1}, \ldots, \varphi_{n+d}$ for factors. Then the development factors $\varphi_1, \ldots, \varphi_n$ form a development pattern for factors and the *tail factors* $\varphi_{n+1}, \ldots, \varphi_{n+d}$ form a *tail pattern for factors*.

All methods that are based on a development pattern provide estimators of the development factors $\varphi_1, \ldots, \varphi_n$, and we assume that such estimators

$$\widehat{\varphi}_1, \ldots, \widehat{\varphi}_n$$

are given. In the following we also refer to these estimators as *a priori estimators*.

For the estimation of the parameters $\varphi_{n+1}, \ldots, \varphi_{n+d}$ of the tail pattern, it is appropriate to choose a parametric class Φ of functions $\varphi : \mathbb{N} \to (0, \infty)$ and to determine a function $\varphi^* \in \Phi$ that minimizes the *squared approximation error*

$$\sum_{k=1}^{n} \left(\varphi(k) - \widehat{\varphi}_k \right)^2$$

over all functions $\varphi \in \Phi$. Then

$$\varphi^*(1), \ldots, \varphi^*(n)$$

are new estimators of the development factors $\varphi_1, \ldots, \varphi_n$ and

$$\varphi^*(n+1), \ldots, \varphi^*(n+d)$$

are estimators of the tail factors $\varphi_{n+1}, \ldots, \varphi_{n+d}$.

Example. In practice one considers, among others, the classes of all functions φ given by

- (*Sherman curve*)

$$\varphi(k) := 1 + a\,(b+k)^{-c}$$

with $a, c > 0$ and $b \geq 0$.
- (*exponential curve*)

$$\varphi(k) := 1 + a\,\exp(-dk)$$

with $a, d > 0$.
- (*Hoerl curve*)

$$\varphi(k) := a\,(b+k)^c\,\exp(dk)$$

with $a > 0$, $b \geq 0$ and $c, d \in \mathbb{R}$, or
- (*Weibull curve*)

$$\varphi(k) := \frac{d}{1 - \exp(-a\,(b+k)^c)}$$

with $a, c, d > 0$ and $b \geq 0$.

The Sherman curve includes as special cases the *Pareto curve* ($a = 1$) and the *inverse power curve* ($b = 0$). In general, it is often sufficient and even reasonable to consider smaller classes of functions, for example by selecting the shift parameter as $b := 0$ or by fixing other parameters, since an improvement of the approximation by curves with a larger number of parameters does not necessarily result in an improvement of the extrapolation.

The a priori estimators $\widehat{\varphi}_1, \ldots, \widehat{\varphi}_n$ can now be replaced by the estimators $\varphi^*(1), \ldots, \varphi^*(n)$, which in general contain more information than the a priori estimators; however, whether this is reasonable or not has to be decided in every particular application. In certain cases, this replacement may lead to a method of loss reserving that is more general than that which has been selected in the beginning:

- Starting with the chain ladder method and hence with $\widehat{\varphi}_k := \varphi_k^{\mathrm{CL}}$, the replacement of φ_k^{CL} with $\varphi^*(k)$ leads to the loss development method with $\gamma^*(k)$ instead of γ_k^{CL}.
- Starting with the additive method and hence with $\widehat{\varphi}_k := \varphi_k^{\mathrm{AD}}$, the replacement of φ_k^{AD} with $\varphi^*(k)$ leads to the Cape Cod method with $\gamma^*(k)$ instead of γ_k^{AD}.

Here $\gamma^*(0), \gamma^*(1), \ldots, \gamma^*(n+d)$ with $\gamma^*(n+d) = 1$ are the estimators of the extended development pattern for quotas which are obtained by conversion from the estimators $\varphi^*(1), \ldots, \varphi^*(n+d)$ of the extended development pattern for factors. Thus, in both cases the replacement of estimators leads to a more general method, which however is still subject to the Bornhuetter–Ferguson principle like the original method.

Furthermore, in the minimization of the squared approximation error, it may be reasonable to exclude the first or also the last development years when the respective a priori estimators show particularities.

Extended Development Pattern for Quotas

Given a development pattern for factors $\varphi_1, \ldots, \varphi_n$, there also exists a development pattern for quotas $\gamma_0, \gamma_1, \ldots, \gamma_n$ with $\gamma_n = 1$; correspondingly, given an extended development pattern for factors $\varphi_1, \ldots, \varphi_n, \varphi_{n+1}, \ldots, \varphi_{n+d}$, there also exists an *extended development pattern for quotas* $\gamma_0', \gamma_1', \ldots, \gamma_n', \gamma_{n+1}', \ldots, \gamma_{n+d}'$ with $\gamma_{n+d}' = 1$. For all $k \in \{0, 1, \ldots, n\}$, the two development pattern for quotas are related by the identity

$$\gamma_k' = \frac{E[S_{i,k}]}{E[S_{i,n+d}]} = \frac{E[S_{i,k}]}{E[S_{i,n}]} \frac{E[S_{i,n}]}{E[S_{i,n+d}]} = \gamma_k \, \gamma_n'$$

which can also be written as

$$\frac{\gamma_k'}{\gamma_n'} = \gamma_k$$

Similarly, given a priori estimators $\widehat{\varphi}_1, \ldots, \widehat{\varphi}_n$ of the development pattern for factors $\varphi_1, \ldots, \varphi_n$, there also exist a priori estimators $\widehat{\gamma}_0, \widehat{\gamma}_1, \ldots, \widehat{\gamma}_n$ with $\widehat{\gamma}_n = 1$ of the development pattern for quotas $\gamma_0, \gamma_1, \ldots, \gamma_n$. To estimate the parameters of the extended development pattern for quotas $\gamma_0', \gamma_1', \ldots, \gamma_n', \gamma_{n+1}', \ldots, \gamma_{n+d}'$, it is then appropriate to choose, as before, a parametric class Γ of functions $\gamma : \mathbb{N}_0 \to (0, \infty)$ and to determine a function $\gamma^* \in \Gamma$ that minimizes the *squared approximation error*

$$\sum_{k=0}^{n} \left(\frac{\gamma(k)}{\gamma(n)} - \widehat{\gamma}_k \right)^2$$

over all functions $\gamma \in \Gamma$. Then

$$\frac{\gamma^*(0)}{\gamma^*(n+d)}, \quad \frac{\gamma^*(1)}{\gamma^*(n+d)}, \quad \dots, \quad \frac{\gamma^*(n)}{\gamma^*(n+d)}, \quad \frac{\gamma^*(n+1)}{\gamma^*(n+d)}, \quad \dots, \quad \frac{\gamma^*(n+d)}{\gamma^*(n+d)} = 1$$

are estimators of the extended development pattern for quotas.

In the case of claim numbers or paid losses it is natural to choose Γ as a parametric class of distribution functions of distributions on $\mathbb{N}_0$.

Example. In practice the class of all functions γ with

$$\gamma(k) := p \left(1 - \exp(-d_1 k) \right) + (1-p) \left(1 - \exp(-d_2 k) \right)$$

(*mixed exponential curve*) with $d_1, d_2 > 0$ and $p \in [0, 1]$ is used among others.

In the case of incurred losses, however, this approach may be inappropriate because of the possibility of negative expected incremental losses.

Extended Development Pattern for Incremental Quotas

In principle, the methods for estimating the parameters of an extended development pattern for factors or quotas presented so far can also be used to estimate an extended development pattern for incremental quotas. Since a priori estimators of factors and quotas tend to contain more information than those of incremental quotas and since they also provide a smoothed representation of the development of losses, estimation of the parameters of an extended development pattern for incremental quotas is presumably only of marginal interest.

Infinite Development Pattern for Incremental Quotas

A general approach for choosing the classes Φ and Γ consists in starting with a parametric class Θ of probability distributions on $\mathbb{N}_0$. In particular, a class Θ of functions $\vartheta : \mathbb{N}_0 \to (0, 1)$ satisfying $\sum_{k=0}^{\infty} \vartheta(k) = 1$ yields a parametric class Γ of monotonically increasing functions $\gamma : \mathbb{N}_0 \to (0, 1)$ given by

$$\gamma(k) := \sum_{l=0}^{k} \vartheta(l)$$

and a parametric class Φ of functions $\varphi : \mathbb{N} \to (1, \infty)$ given by

$$\varphi(k) := \frac{\gamma(k)}{\gamma(k-1)}$$

The values of the functions ϑ, γ, φ can be understood as *infinite development patterns* for incremental quotas, quotas and factors, respectively.

Example. Consider the parametric class Θ of all probability distributions $\vartheta_\eta : \mathbb{N}_0 \to (0, 1)$ given by

$$\vartheta_\eta(k) := (1-\eta)\, \eta^k$$

for some $\eta \in (0, 1)$ and all $k \in \mathbb{N}_0$. Then the functions $\gamma_\eta : \mathbb{N}_0 \to (0, 1)$ given by

$$\gamma_\eta(k) := 1 - \eta^{k+1}$$

form a parametric class Γ of monotonically increasing functions and the functions $\varphi_\eta : \mathbb{N} \to (1, \infty)$ given by

$$\varphi_\eta(k) := \frac{1 - \eta^{k+1}}{1 - \eta^k}$$

form a parametric class Φ of monotonically decreasing functions.

Tail Length

So far, the *tail length d* was assumed as given, but it is hard to determine in practice.

The tail length results primarily from the characteristics of the portfolio, for example from the maximal duration of court proceedings or from the maximal payment period of annuities.

In the approximation of the a priori estimators of the parameters of the extended development pattern for quotas, the tail length d could be determined such that

$$\varphi^*(n+d) \le 1 + \varepsilon$$

with, for example, $\varepsilon := 0.001$.

Notes

Keywords: Additive Method, Bornhuetter–Ferguson Method, Bornhuetter–Ferguson Principle, Cape Cod Method, Chain Ladder Method (Basics), Development Patterns (Basics), Development Patterns (Estimation), Loss Development Method, Reinsurance, Run-Off Data, Run-Off Triangles.

References: GDV [2011], Lowe & Mohrman [1985], Sherman [1984, 1985].

Volume Measures

Michael Radtke and Klaus D. Schmidt

Many methods of loss reserving use known *volume measures* $v_0, v_1, \ldots, v_n$ of the accident years. Possible volume measures v_i are

- the number of contracts,
- the sum insured,
- the limit of liability or the premium calculation basis (in *liability insurance*),
- the written, accounted or earned premium or
- the expected number of claims
  of the portfolio in accident year $i \in \{0, 1, \ldots, n\}$.

For given volume measures $v_0, v_1, \ldots, v_n$, the ratio

$$\frac{S_{i,n}}{v_i}$$

is called the *ultimate loss ratio* of accident year i and the ratio

$$\frac{Z_{i,k}}{v_i}$$

is called the *incremental loss ratio* of accident year i and development year k. Then

$$\kappa_i := E\left[\frac{S_{i,n}}{v_i}\right]$$

M. Radtke (✉)
Fachhochschule Dortmund, Dortmund, Germany
e-mail: michael.radtke@fh-dortmund.de

K.D. Schmidt
Technische Universität Dresden, Dresden, Germany
e-mail: klaus.d.schmidt@tu-dresden.de

© Springer International Publishing Switzerland 2016
M. Radtke et al. (eds.), *Handbook on Loss Reserving*,
EAA Series, DOI 10.1007/978-3-319-30056-6_38

is the *expected ultimate loss ratio* of accident year i and

$$\zeta_{i,k} := E\left[\frac{Z_{i,k}}{v_i}\right]$$

is the *expected incremental loss ratio* of accident year i and development year k. In certain models of loss reserving it is assumed that the expected ultimate loss ratios or the expected incremental loss ratios of a given development year are identical for all accident years.

Notes

Keywords: Additive Method, Bornhuetter–Ferguson Method, Cape Cod Method, Linear Models (Loss Reserving), Loss Ratios, Reinsurance, Separation Method.

Probability Distributions

Klaus D. Schmidt

For a random variable X, the mapping $F_X : \mathbb{R} \to [0, 1]$ given by

$$F_X(x) := P[\{X \le x\}]$$

is called the *distribution function* of X. The distribution function F_X describes the stochastic properties of the random variable X. Using the distribution function one obtains, for example,

$$P[\{x_1 < X \le x_2\}] = F_X(x_2) - F_X(x_1)$$

and

$$P[\{x < X\}] = 1 - F_X(x)$$

This means that the distribution function of a random variable determines its *probability distribution* (*distribution* for short).

A random variable X has a *discrete distribution* if there exists a countable set $Q \subseteq \mathbb{R}$ such that $P[\{X \in Q\}] = 1$. In this case the distribution function of X satisfies

$$F_X(x) = \sum_{t \in Q \cap (-\infty, x]} P[\{X = t\}]$$

for all $x \in \mathbb{R}$. The case $Q \subseteq \mathbb{N}_0$ or even $Q = \mathbb{N}_0$ is of particular interest.

K.D. Schmidt (✉)
Technische Universität Dresden, Dresden, Germany
e-mail: klaus.d.schmidt@tu-dresden.de

© Springer International Publishing Switzerland 2016
M. Radtke et al. (eds.), *Handbook on Loss Reserving*,
EAA Series, DOI 10.1007/978-3-319-30056-6_39

A random variable X has a *continuous distribution* if there exists a function $f_X : \mathbb{R} \to \mathbb{R}_+$ such that

$$F_X(x) = \int_{-\infty}^{x} f_X(t)\, dt$$

holds for all $x \in \mathbb{R}$ and in this case the function f_X is called the *density function* of X.

The representations of a distribution function in terms of sums or integrals can only in exceptional cases be replaced by a representation in closed form. They are nevertheless sufficient to compute the *expectation* $E[X]$ of a random variable X by means of the identity

$$E[X] = \sum_{x \in Q} x\, P[\{X = x\}]$$

in the case of a discrete distribution and by

$$E[X] = \int_{-\infty}^{\infty} x\, f_X(x)\, dx$$

in the case of a continuous distribution, provided in both cases that the expectation exists. In both cases the *variance* var$[X]$ of X is defined by

$$\mathrm{var}[X] := E\big[(X - E[X])^2\big] = E[X^2] - \big(E[X]\big)^2$$

The square root of the variance of X is called the *standard deviation* of X.

For a random variable X and $\alpha \in (0, 1)$, every $q \in \mathbb{R}$ satisfying

$$P[\{X \le q\}] \ge \alpha \quad \text{and} \quad P[\{X \ge q\}] \ge 1 - \alpha$$

is called the *α-quantile* or briefly a *quantile* of the distribution of X. The quantiles of the distribution of X may fail to be unique. This applies, in particular, when X has a discrete distribution. By contrast, all quantiles are unique when the distribution function of X is strictly monotone on the interval $\{x \in \mathbb{R} \mid 0 < F_X(x) < 1\}$.

The distribution function $F_{\mathbf{X}} : \mathbb{R}^m \to [0, 1]$ of a random vector $\mathbf{X}$ with values in $\mathbb{R}^m$ is defined as in the case of a random variable and also the notions of a discrete or continuous distribution are extended in the obvious way from random variables to random vectors. The *expectation* $E[\mathbf{X}]$ of the random vector $\mathbf{X}$ is defined as the vector

$$E[\mathbf{X}] := \begin{pmatrix} E[X_1] \\ E[X_2] \\ \vdots \\ E[X_m] \end{pmatrix}$$

consisting of the expectations of the coordinates of $\mathbf{X}$ and the *variance* var[$\mathbf{X}$] of $\mathbf{X}$ is defined as the matrix

$$\mathrm{var}[\mathbf{X}] := \begin{pmatrix} \mathrm{cov}[X_1, X_1] & \mathrm{cov}[X_1, X_2] & \cdots & \mathrm{cov}[X_1, X_m] \\ \mathrm{cov}[X_2, X_1] & \mathrm{cov}[X_2, X_2] & \cdots & \mathrm{cov}[X_2, X_m] \\ \vdots & \vdots & & \vdots \\ \mathrm{cov}[X_m, X_1] & \mathrm{cov}[X_m, X_2] & \cdots & \mathrm{cov}[X_m, X_m] \end{pmatrix}$$

consisting of the *covariances*

$$\mathrm{cov}[X_i, X_j] := E\big[(X_i - E[X_i])(X_j - E[X_j])\big] = E[X_i X_j] - E[X_i]\,E[X_j]$$

of the coordinates of $\mathbf{X}$. Note that $\mathrm{cov}[X_i, X_i] = \mathrm{var}[X_i]$. Since the covariance of any two random variables is symmetric, the matrix var[$\mathbf{X}$] is symmetric as well. Moreover, normalization of the covariances by means of the standard deviations yields the *coefficients of correlation*

$$\varrho[X_i, X_j] := \frac{\mathrm{cov}[X_i, X_j]}{\sqrt{\mathrm{var}[X_i]}\,\sqrt{\mathrm{var}[X_j]}} = \mathrm{cov}\left[\frac{X_i - E[X_i]}{\sqrt{\mathrm{var}[X_i]}}, \frac{X_j - E[X_j]}{\sqrt{\mathrm{var}[X_j]}}\right]$$

and one has $\varrho[X_i, X_j] \in [-1, 1]$.

Beta Distribution

A random variable X has the *Beta distribution*

$$\mathbf{Be}(\alpha, \beta)$$

with parameters $\alpha, \beta \in (0, \infty)$ if

$$P[\{X \le x\}] = \int_{-\infty}^{x} \frac{\Gamma(\alpha+\beta)}{\Gamma(\alpha)\,\Gamma(\beta)}\, t^{\alpha-1}\,(1-t)^{\beta-1}\,\chi_{(0,1)}(t)\,dt$$

holds for all $x \in \mathbb{R}$. In this case one has

$$E[X] = \frac{\alpha}{\alpha + \beta}$$

$$\mathrm{var}[X] = \frac{\alpha\beta}{(\alpha+\beta)^2(\alpha+\beta+1)}$$

In particular, $\mathbf{Be}(1, 1) = \mathbf{Uni}(0, 1)$.

Binomial Distribution

A random variable X has the *binomial distribution*

$$\mathbf{B}(n; \vartheta)$$

with parameters $n \in \mathbb{N}$ and $\vartheta \in (0, 1)$ if

$$P[\{X = k\}] = \binom{n}{k}\vartheta^k(1-\vartheta)^{n-k}$$

holds for all $k \in \{0, 1, \ldots, n\}$. In this case one has

$$E[X] = n\,\vartheta$$
$$\mathrm{var}[X] = n\,\vartheta\,(1-\vartheta)$$

and hence $\mathrm{var}[X] < E[X]$.

Dirichlet Distribution

A random vector $\mathbf{X}$ with values in $\mathbb{R}^m$ and

$$P\left[\left\{\sum_{i=1}^{m} X_i = 1\right\}\right] = 1$$

has the *Dirichlet distribution*

$$\mathbf{Dir}(\eta_1, \ldots, \eta_m)$$

with parameters $\eta_1, \ldots, \eta_m \in (0, \infty)$ if

$$P[\{\mathbf{Z} \leq \mathbf{z}\}] = \int_{-\infty}^{\mathbf{z}} \frac{\Gamma(\sum_{i=1}^{m} \eta_i)}{\prod_{i=1}^{m} \Gamma(\eta_i)} \left(\prod_{i=1}^{m-1} t_i^{\eta_i - 1} \right) \left(1 - \sum_{i=1}^{m-1} t_i \right)^{\eta_m - 1} \chi_{S^{m-1}}(\mathbf{t}) \, d\mathbf{t}$$

holds for all $\mathbf{z} \in \mathbb{R}^{m-1}$; here $\mathbf{Z}$ denotes the random vector with values in $\mathbb{R}^{m-1}$ which results from the random vector $\mathbf{X}$ by deleting its last coordinate and the set S^{m-1} is defined by

$$S^{m-1} := \left\{ \mathbf{t} \in (0, \infty)^{m-1} \;\middle|\; \sum_{i=1}^{m-1} t_i < 1 \right\}$$

In this case one has, for all $i, j \in \{1, \ldots, m\}$,

$$E[X_i] = \frac{\eta_i}{\eta}$$

$$E[X_i X_j] = \frac{\eta_i}{\eta \, (\eta + 1)} \, \delta_{i,j} + \frac{\eta_i \, \eta_j}{\eta \, (\eta + 1)}$$

where $\eta := \sum_{i=1}^{m} \eta_i$. In particular, $\mathbf{Dir}(\eta_1, \eta_2) = \mathbf{Be}(\eta_1, \eta_2)$.

Exponential Distribution

A random variable X has the *exponential distribution*

$$\mathbf{Exp}(\alpha)$$

with parameter $\alpha \in (0, \infty)$ if

$$P[\{X \leq x\}] = \int_{-\infty}^{x} \alpha \, e^{-\alpha t} \, \chi_{(0,\infty)}(t) \, dt = \begin{cases} 0 & \text{if } x \leq 0 \\ 1 - e^{-\alpha x} & \text{if } x > 0 \end{cases}$$

holds for all $x \in \mathbb{R}$. In this case one has

$$E[X] = 1/\alpha$$
$$\mathrm{var}[X] = 1/\alpha^2$$

In particular, $\mathbf{Exp}(\alpha) = \mathbf{Ga}(\alpha, 1)$.

Gamma Distribution

A random variable X has the *Gamma distribution*

$$\mathbf{Ga}(\alpha, \beta)$$

with parameters $\alpha, \beta \in (0, \infty)$ if

$$P[\{X \leq x\}] = \int_{-\infty}^{x} \frac{\alpha^{\beta}}{\Gamma(\beta)} e^{-\alpha t} t^{\beta-1} \chi_{(0,\infty)}(t) \, dt$$

holds for all $x \in \mathbb{R}$. In this case one has

$$E[X] = \beta/\alpha$$
$$\mathrm{var}[X] = \beta/\alpha^2$$

In particular, $\mathbf{Ga}(\alpha, 1) = \mathbf{Exp}(\alpha)$.

Lognormal Distribution

A random vector $\mathbf{X}$ with values in $\mathbb{R}^m$ has the *lognormal distribution*

$$\mathbf{LN}(\boldsymbol{\mu}, \Sigma)$$

with $\boldsymbol{\mu} \in \mathbb{R}^m$ and a positive definite symmetric matrix $\Sigma \in \mathbb{R}^{m \times m}$ if

$$P[\{\mathbf{X} \leq \mathbf{x}\}] = \int_{-\infty}^{\mathbf{x}} \frac{1}{\sqrt{(2\pi)^m \det(\Sigma)} \prod_{i=1}^{m} t_i}$$
$$\exp\left(-\frac{1}{2}(\ln(\mathbf{t}) - \boldsymbol{\mu})' \Sigma^{-1} (\ln(\mathbf{t}) - \boldsymbol{\mu})\right) \chi_{(0,\infty)^m}(\mathbf{t}) \, d\mathbf{t}$$

holds for all $\mathbf{x} \in \mathbb{R}^m$. In this case one has

$$E[X_i] = \exp\left(\mathbf{e}_i' \boldsymbol{\mu} + \frac{1}{2} \mathbf{e}_i' \Sigma \mathbf{e}_i\right)$$

$$E[X_i X_j] = \exp\left((\mathbf{e}_i + \mathbf{e}_j)' \boldsymbol{\mu} + \frac{1}{2} (\mathbf{e}_i + \mathbf{e}_j)' \Sigma (\mathbf{e}_i + \mathbf{e}_j)\right)$$

$$\mathrm{cov}[X_i, X_j] = E[X_i] E[X_j] \left(\exp(\mathbf{e}_i' \Sigma \mathbf{e}_j) - 1\right)$$

The random vector $\mathbf{X}$ has the lognormal distribution $\mathbf{LN}(\boldsymbol{\mu}, \Sigma)$ if and only if the random vector $\ln(\mathbf{X})$ has the normal distribution $\mathbf{N}(\boldsymbol{\mu}, \Sigma)$.

Multinomial Distribution

A random vector $\mathbf{X}$ with values in $\mathbb{R}^m$ has the *multinomial distribution*

$$\mathbf{M}(n; \vartheta_1, \ldots, \vartheta_m)$$

with parameters $n \in \mathbb{N}_0$ and $\vartheta_1, \ldots, \vartheta_m \in (0, 1)$ satisfying $\sum_{i=1}^{m} \vartheta_i = 1$ if

$$P[\{\mathbf{X} = \mathbf{k}\}] = \frac{n!}{\prod_{i=1}^{m} k_i!} \prod_{i=1}^{m} \vartheta_i^{k_i}$$

holds for all $\mathbf{k} \in \mathbb{N}_0^m$ such that $\sum_{i=1}^{m} k_i = n$. In this case one has

$$P\left[\left\{\sum_{i=1}^{m} X_i = n\right\}\right] = 1$$

and for all $i, j \in \{1, \ldots, m\}$ one has

$$E[X_i] = n\,\vartheta_i$$
$$\mathrm{cov}[X_i, X_j] = n\,\vartheta_i\,\delta_{i,j} - n\,\vartheta_i\,\vartheta_j$$

In particular, $\mathbf{M}(n; \vartheta, 1-\vartheta) = \mathbf{B}(n; \vartheta)$.

Negative Binomial Distribution

A random variable X has the *negative binomial distribution*

$$\mathbf{NB}(\alpha, \vartheta)$$

with parameters $\alpha \in (0, \infty)$ and $\vartheta \in (0, 1)$ if

$$P[\{X = k\}] = \binom{\alpha + k - 1}{k}(1-\vartheta)^\alpha\,\vartheta^k$$

holds for all $k \in \mathbb{N}_0$. In this case one has

$$E[X] = \frac{\alpha\,\vartheta}{1 - \vartheta}$$
$$\mathrm{var}[X] = \frac{\alpha\,\vartheta}{(1-\vartheta)^2}$$

and hence $\text{var}[X] > E[X]$. Since

$$P[\{X = k\}] = \binom{\alpha \mid k-1}{k}(1-\vartheta)^\alpha\,\vartheta^k$$

$$= \int_{-\infty}^{\infty} e^{-\lambda}\,\frac{\lambda^k}{k!}\,\frac{((1-\vartheta)/\vartheta)^\alpha}{\Gamma(\alpha)}\,e^{-((1-\vartheta)/\vartheta)\lambda}\,\lambda^{\alpha-1}\,\chi_{(0,\infty)}(\lambda)\,d\lambda$$

every negative binomial distribution is a mixture of all Poisson distributions under a Gamma distribution.

Normal Distribution

A random vector $\mathbf{X}$ with values in $\mathbb{R}^m$ has the *normal distribution*

$$\mathbf{N}(\mu, \Sigma)$$

with $\mu \in \mathbb{R}^m$ and a positive definite symmetric matrix $\Sigma \in \mathbb{R}^{m \times m}$ if

$$P[\{\mathbf{X} \leq \mathbf{x}\}] = \int_{-\infty}^{\mathbf{x}} \frac{1}{\sqrt{(2\pi)^m \det(\Sigma)}}\,\exp\left(-\frac{1}{2}(\mathbf{t}-\mu)'\Sigma^{-1}(\mathbf{t}-\mu)\right) d\mathbf{t}$$

holds for all $\mathbf{x} \in \mathbb{R}^m$. In this case one has

$$E[\mathbf{X}] = \mu$$
$$\text{var}[\mathbf{X}] = \Sigma$$

If the random vector $\mathbf{X}$ has the normal distribution $\mathbf{N}(\mu, \Sigma)$, then the random vector $\mathbf{d} + \mathbf{D}\mathbf{X}$ with a vector $\mathbf{d} \in \mathbb{R}^s$ and a matrix $\mathbf{D} \in \mathbb{R}^{s \times m}$ has the normal distribution $\mathbf{N}(\mathbf{d}+\mathbf{D}\mu, \mathbf{D}\Sigma\mathbf{D}')$.

Pareto Distribution

A random variable X has the *Pareto distribution*

$$\mathbf{Par}(\alpha, \beta)$$

with parameters $\alpha, \beta \in (0, \infty)$ if

$$P[\{X \leq x\}] = \int_{-\infty}^{x} \frac{\beta}{\alpha} \left(\frac{\alpha}{t}\right)^{\beta+1} \chi_{(\alpha,\infty)}(t) \, dt = \begin{cases} 0 & \text{if } x \leq \alpha \\ 1 - \left(\dfrac{\alpha}{x}\right)^{\beta} & \text{if } x > \alpha \end{cases}$$

holds for all $x \in \mathbb{R}$. In the case $\beta \in (1, \infty)$ one has

$$E[X] = \frac{\alpha \beta}{\beta - 1}$$

and in the case $\beta \in (2, \infty)$ one has

$$\text{var}[X] = \frac{\alpha^2 \beta}{(\beta-1)^2(\beta-2)}$$

It thus depends on the value of the parameter β whether the expectation or the variance of X is finite or not.

Poisson Distribution

A random variable X has the *Poisson distribution*

$$\mathbf{Poi}(\alpha)$$

with parameter $\alpha \in (0, \infty)$ if

$$P[\{X = k\}] = e^{-\alpha} \frac{\alpha^k}{k!}$$

holds for all $k \in \mathbb{N}_0$. In this case one has

$$E[X] = \alpha$$
$$\text{var}[X] = \alpha$$

and hence $\text{var}[X] = E[X]$.

Uniform Distribution

A random variable X has the *uniform distribution*

$$\textbf{Uni}(0, 1)$$

if

$$P[\{X \leq x\}] = \int_{-\infty}^{x} \chi_{(0,1)}(t)\, dt = \begin{cases} 0 & \text{if} \quad x \leq 0 \\ x & \text{if} \quad 0 < x \leq 1 \\ 1 & \text{if} \quad 1 < x \end{cases}$$

holds for all $x \in \mathbb{R}$. In this case one has

$$E[X] = 1/2$$
$$\text{var}[X] = 1/12$$

In particular, $\textbf{Uni}(0, 1) = \textbf{Be}(1, 1)$.

Notes

References: Müller [1991], Schmidt [2009, 2011].

List of Symbols

Run-Off Quantities

$N_{i,k}$ *incremental claim number* from accident year i in development year k

$S_{i,k}$ *cumulative loss* from accident year i in development year k

$$S_{i,k} = \sum_{l=0}^{k} Z_{i,l}$$

$S_{i,n-i}$ *current loss* from accident year i

$S_{i,n}$ *ultimate loss* from accident year i

$Z_{i,k}$ *incremental loss* from accident year i in development year k

$$Z_{i,k} = \begin{cases} S_{i,0} & \text{if } k = 0 \\ S_{i,k} - S_{i,k-1} & \text{else} \end{cases}$$

Parameters

v_i *volume measure* of accident year i (known)

α_i *expected ultimate loss* of accident year i

$$\alpha_i = E[S_{i,n}]$$

κ_i *expected ultimate loss ratio* of accident year i

$$\kappa_i = E[S_{i,n}/v_i]$$

© Springer International Publishing Switzerland 2016
M. Radtke et al. (eds.), *Handbook on Loss Reserving*,
EAA Series, DOI 10.1007/978-3-319-30056-6

ϑ_k *incremental quota* of development year k with

$$\vartheta_k = E[Z_{i,k}]/E[S_{i,n}]$$

for all $i \in \{0, 1, \ldots, n\}$

γ_k *quota* of development year k with

$$\gamma_k = E[S_{i,k}]/E[S_{i,n}]$$

for all $i \in \{0, 1, \ldots, n\}$

φ_k *development factor* of development year k with

$$\varphi_k = E[S_{i,k}]/E[S_{i,k-1}]$$

for all $i \in \{0, 1, \ldots, n\}$

ξ_k *incremental development factor* of development year k with

$$\xi_k = E[Z_{i,k}]/E[Z_{i,0}]$$

for all $i \in \{0, 1, \ldots, n\}$

ζ_k *expected incremental loss ratio* of development year k with

$$\zeta_k = E[Z_{i,k}]/v_i$$

for all $i \in \{0, 1, \ldots, n\}$

κ *expected ultimate loss ratio* with

$$\kappa = E[S_{i,n}]/v_i$$

for all $i \in \{0, 1, \ldots, n\}$

Estimators of the Parameters

$\widehat{\vartheta}_{i,k}$ *individual incremental quota* of development year k and accident year i
$$\widehat{\vartheta}_{i,k} = Z_{i,k}/S_{i,n}$$

$\widehat{\gamma}_{i,k}$ *individual quota* of development year k and accident year i
$$\widehat{\gamma}_{i,k} = S_{i,k}/S_{i,n}$$

$\widehat{\varphi}_{i,k}$ *individual development factor* of development year k and accident year i
$$\widehat{\varphi}_{i,k} = S_{i,k}/S_{i,k-1}$$

$\widehat{\xi}_{i,k}$ *individual incremental development factor* of development year k and accident year i

$$\widehat{\xi}_{i,k} = Z_{i,k}/Z_{i,0}$$

$\widehat{\zeta}_{i,k}$ *individual incremental loss ratio* of development year k and accident year i

$$\widehat{\zeta}_{i,k} = Z_{i,k}/v_i$$

$\widehat{\kappa}_{i,k}$ *individual ultimate loss ratio* of development year k and accident year i

$$\widehat{\kappa}_{i,k} = S_{i,k}/(v_i\widehat{\gamma}_k)$$

with estimators $\widehat{\gamma}_0, \widehat{\gamma}_1, \ldots, \widehat{\gamma}_n$ of the quotas $\gamma_0, \gamma_1, \ldots, \gamma_n$

$\widehat{\kappa}_i$ *individual ultimate loss ratio* of accident year i

$$\widehat{\kappa}_i = S_{i,n-i}/(v_i\widehat{\gamma}_{n-i}) = \widehat{\kappa}_{i,n-i}$$

with estimators $\widehat{\gamma}_0, \widehat{\gamma}_1, \ldots, \widehat{\gamma}_n$ of the quotas $\gamma_0, \gamma_1, \ldots, \gamma_n$

φ_k^{CL} *chain ladder factor* of development year k

$$\varphi_k^{\mathrm{CL}} = \frac{\sum_{j=0}^{n-k} S_{j,k}}{\sum_{j=0}^{n-k} S_{j,k-1}} = \sum_{j=0}^{n-k} \frac{S_{j,k-1}}{\sum_{h=0}^{n-k} S_{h,k-1}} \widehat{\varphi}_{j,k}$$

ξ_k^{PA} *Panning factor* of development year k

$$\xi_k^{\mathrm{PA}} = \frac{\sum_{j=0}^{n-k} Z_{j,0} Z_{j,k}}{\sum_{j=0}^{n-k} Z_{j,0}^2} = \sum_{j=0}^{n-k} \frac{Z_{j,0}^2}{\sum_{h=0}^{n-k} Z_{h,0}^2} \widehat{\xi}_{j,k}$$

ζ_k^{AD} *additive incremental loss ratio* of development year k

$$\zeta_k^{\mathrm{AD}} = \frac{\sum_{j=0}^{n-k} Z_{j,k}}{\sum_{j=0}^{n-k} v_j} = \sum_{j=0}^{n-k} \frac{v_j}{\sum_{h=0}^{n-k} v_h} \widehat{\zeta}_{j,k}$$

κ^{AD} *additive ultimate loss ratio*

$$\kappa^{\mathrm{AD}} = \sum_{l=0}^{n} \zeta_l^{\mathrm{AD}}$$

κ^{CC} *Cape Cod ultimate loss ratio*

$$\kappa^{\mathrm{CC}} = \frac{\sum_{j=0}^n S_{J,n-j}}{\sum_{j=0}^n v_j \widehat{\gamma}_{n-j}} = \sum_{j=0}^n \frac{v_j \widehat{\gamma}_{n-j}}{\sum_{h=0}^n v_h \widehat{\gamma}_{n-h}} \widehat{\kappa}_j$$

with estimators $\widehat{\gamma}_0, \widehat{\gamma}_1, \dots, \widehat{\gamma}_n$ of the quotas $\gamma_0, \gamma_1, \dots, \gamma_n$

Classes of Estimators and Predictors

AD	additive method
BF	Bornhuetter–Ferguson method
CC	Cape Cod method
CL	chain ladder method
CR	credibility
EL	expected loss method
GM	Gauss–Markov
GU	grossing up method
LD	loss development method
PA	Panning method

Sets of Numbers and Vectors

$\mathbb{N}$	the set $\{1, 2, \dots\}$ of natural numbers (without zero)
$\mathbb{N}_0$	the set $\{0, 1, 2, \dots\}$ of natural numbers (including zero)
$\mathbb{R}$	the set of real numbers
$\mathbb{R}_+$	the set $[0, \infty)$
$\mathbb{R}^m$	the Euclidean space of dimension m

Vectors and Matrices

0   the vector

$$\mathbf{0} := \begin{pmatrix} 0 \\ \vdots \\ 0 \end{pmatrix}$$

of the Euclidean space $\mathbb{R}^m$

$\mathbf{e}_i$ the i-th unit vector of the Euclidean space $\mathbb{R}^m$

1   the vector

$$\mathbf{1} := \begin{pmatrix} 1 \\ \vdots \\ 1 \end{pmatrix} = \sum_{i=1}^m \mathbf{e}_i$$

of the Euclidean space $\mathbb{R}^m$

$\mathbf{x}$	the vector	

$$\mathbf{x} = \begin{pmatrix} x_1 \\ \vdots \\ x_m \end{pmatrix} = \sum_{i=1}^{m} x_i \, \mathbf{e}_i$$

of the Euclidean space $\mathbb{R}^m$

$\mathbf{x}'$	the transpose

$$\mathbf{x}' := \begin{pmatrix} x_1 \ \ldots \ x_m \end{pmatrix}$$

of the vector $\mathbf{x} \in \mathbb{R}^m$

$\ln(\mathbf{x})$	the vector

$$\ln(\mathbf{x}) := \begin{pmatrix} \ln(x_1) \\ \vdots \\ \ln(x_m) \end{pmatrix}$$

with $\mathbf{x} \in (0, \infty)^m$

$\mathbf{O}$	the matrix

$$\mathbf{O} := \begin{pmatrix} 0 \ \ldots \ 0 \\ \vdots \quad \vdots \\ 0 \ \ldots \ 0 \end{pmatrix}$$

$\mathbf{E}$	the matrix

$$\mathbf{E} := \begin{pmatrix} 1 \ \ldots \ 1 \\ \vdots \quad \vdots \\ 1 \ \ldots \ 1 \end{pmatrix} = \mathbf{1}\mathbf{1}'$$

$\mathbf{A}$	the matrix

$$\mathbf{A} = \begin{pmatrix} a_{1,1} \ \ldots \ a_{1,n} \\ \vdots \quad \vdots \\ a_{m,1} \ \ldots \ a_{m,n} \end{pmatrix}$$

$\mathbf{A}'$	the transpose

$$\mathbf{A}' := \begin{pmatrix} a_{1,1} \ \ldots \ a_{m,1} \\ \vdots \quad \vdots \\ a_{1,n} \ \ldots \ a_{m,n} \end{pmatrix}$$

of the matrix $\mathbf{A} \in \mathbb{R}^{m \times n}$

Kronecker Symbol and Indicator Function

$\delta_{i,j}$ the Kronecker symbol defined by

$$\delta_{i,j} := \begin{cases} 1 & \text{if } i = j \\ 0 & \text{else} \end{cases}$$

χ_B the indicator function of a set B defined by

$$\chi_B(x) := \begin{cases} 1 & \text{if } x \in B \\ 0 & \text{else} \end{cases}$$

Probability Distributions

B$(n; \vartheta)$ *binomial distribution*
with parameters $n \in \mathbb{N}$ and $\vartheta \in (0, 1)$

Be(α, β) *beta distribution*
with parameters $\alpha, \beta \in (0, \infty)$

Dir$(\eta_1, \ldots, \eta_m)$ *Dirichlet distribution*
with parameters $\eta_1, \ldots, \eta_m \in (0, \infty)$

Exp(α) *exponential distribution*
with parameter $\alpha \in (0, \infty)$

Ga(α, β) *gamma distribution*
with parameters $\alpha, \beta \in (0, \infty)$

LN$(\boldsymbol{\mu}, \Sigma)$ *lognormal distribution*
with parameters $\boldsymbol{\mu} \in \mathbb{R}^m$ and $\Sigma \in \mathbb{R}^{m \times m}$

M$(n; \vartheta_1, \ldots, \vartheta_m)$ *multinomial distribution*
with parameters $n \in \mathbb{N}_0$ and $\vartheta_1, \ldots, \vartheta_m \in (0, 1)$

N$(\boldsymbol{\mu}, \Sigma)$ *normal distribution*
with parameters $\boldsymbol{\mu} \in \mathbb{R}^m$ and $\Sigma \in \mathbb{R}^{m \times m}$

NB(α, ϑ) *negative binomial distribution*
with parameters $\alpha \in (0, \infty)$ and $\vartheta \in (0, 1)$

Par(α, β) *Pareto distribution*
with parameters $\alpha, \beta \in (0, \infty)$

Poi(α) *Poisson distribution*
with parameter $\alpha \in (0, \infty)$

Uni$(0, 1)$ *uniform distribution*
on the interval $(0, 1)$

References

Ajne, B.: Additivity of chain-ladder projections. ASTIN Bull. **24**, 313–318 (1994)

Barnett, G., Zehnwirth, B., Dubossarski, E.: When can accident years be regarded as development years? Proc. CAS **92**, 239–256 (2005)

Benktander, G.: An approach to credibility in calculating IBNR for casualty excess reinsurance. Actuar. Rev. **3**(2), 7 (1976)

Bornhuetter, R.L., Ferguson, R.E.: The actuary and IBNR. Proc. CAS **59**, 181–195 (1972)

Braun, C.: The prediction error of the chain-ladder method applied to correlated run-off triangles. ASTIN Bull. **34**, 399–423 (2004)

Buchwalder, M., Bühlmann, H., Merz, M., Wüthrich, M.V.: The mean square error of prediction in the chain-ladder reserving method (Mack and Murphy revisited). ASTIN Bull. **36**, 521–542 (2006)

Daykin, C.D., Pentikäinen, T., Pesonen, M.: Practical Risk Theory for Actuaries. Chapman and Hall, New York (1994)

De Vylder, F.E.: Estimation of IBNR claims by credibility theory. Insur. Math. Econ. **1**, 35–40 (1982)

Dietze, S., Riedrich, T., Schmidt, K.D.: On the Solution of Marginal-Sum Equations. Dresdner Schriften zur Versicherungsmathematik 1/2006 (2006)

Dietze, S., Riedrich, T., Schmidt, K.D.: Marginal-sum equations and related fixed-point problems. Nonlinear Anal. Ser. A: Theory Methods Appl. **75**, 6088–6102 (2012)

Doray, L.G.: UMVUE of the IBNR reserve in a lognormal linear regression model. Insur. Math. Econ. **18**, 43–57 (1996)

England, P.D., Verrall, R.J.: Stochastic claims reserving in general insurance. B. Actuar. J. **8**, 443–518 (2002)

Fuchs, S.: Consistent loss prediction of a portfolio and its subportfolios. Scand. Actuar. J. **6**, 561–581 (2014)

Fuchs, S., Ludwig, A., Schmidt, K.D.: Zur Exaktheit der Standardformel. Z. Gesamte Versicherungswissenschaft **102**, 87–95 (2013)

GDV: Methoden zur Schätzung von Schaden- und Prämienrückstellungen in der Kompositversicherung. Revised version. Berlin, Gesamtverband der Deutschen Versicherungswirtschaft (2011)

Hachemeister, C.A., Stanard, J.N.: IBNR claims count estimation with static lag functions (1975) Unpublished

Halliwell, L.J.: Conjoint prediction of paid and incurred losses. CAS Forum Summer **1997**(1), 241–379 (1997)

© Springer International Publishing Switzerland 2016
M. Radtke et al. (eds.), *Handbook on Loss Reserving*,
EAA Series, DOI 10.1007/978-3-319-30056-6

310

Hamer, M.D.: Loss prediction by generalized least squares - Discussion of Halliwell (1996). Proc. CAS **86**, 748–763 (1999)

Hess, K.T.: Marginal-sum and maximum-likelihood estimation in a multiplicative tariff. ASTA Adv. Stat. Anal. **93**, 221–233 (2009)

Hess, K.T.: Maximum-likelihood and marginal-sum estimation in some particular collective models. ASTA Adv. Stat. Anal. **96**, 311–326 (2012)

Hess, K.T., Liewald, A., Schmidt, K.D.: An extension of Panjer's recursion. ASTIN Bull. **32**, 283–297 (2002)

Hess, K.T., Schmidt, K.D.: Credibility-Modelle in Tarifierung und Reservierung. Allg. Statist. Archiv **85**, 225–246 (2001)

Hess, K.T., Schmidt, K.D.: A comparison of models for the chain-ladder method. Insur. Math. Econ. **31**, 351–364 (2002)

Hess, K.T., Schmidt, K.D., Zocher, M.: Multivariate loss prediction in the multivariate additive model. Insur. Math. Econ. **39**, 185–191 (2006)

Hesselager, O., Witting, T.: A credibility model with random fluctuations in delay probabilities for the prediction of IBNR claims. ASTIN Bull. **18**, 79–90 (1988)

Hossack, I.B., Pollard, J.H., Zehnwirth, B.: Introductory Statistics with Applications in General Insurance, 2nd edn. Cambridge University Press, New York (1999)

Hovinen, E.: Additive and continuous IBNR. ASTIN Colloquium Loen (1981)

Institute of Actuaries, (ed.): Claims Reserving Manual. The Institute of Actuaries, London (1989)

Kalos, M.H., Whitelock, P.A.: Monte Carlo Methods. Wiley, New York (2008)

Klemmt, H.J.: Separierung von Abwicklungsdreiecken nach Basisschäden und Großschäden. Blätter DGVFM **27**, 49–58 (2005)

Kloberdanz, K., Schmidt, K.D.: Prediction in the linear model under a linear constraint. ASTA Adv. Stat. Anal. **92**, 207–215 (2008)

Kloberdanz, K., Schmidt, K.D.: Loss prediction in a linear model under a linear constraint. ASTA Adv. Stat. Anal. **93**, 205–220 (2009)

Kremer, E.: The correlated chain-ladder method for reserving in case of correlated claims developments. Blätter DGVFM **27**, 315–322 (2005)

Lorenz, H., Schmidt, K.D.: Grossing-up, chain-ladder and marginal-sum estimation. Blätter DGVM **24**, 195–200 (1999)

Lowe, S.P., Mohrman, D.F.: Extrapolating, smoothing and interpolating development factors - Discussion of Sherman (1984). Proc. CAS **72**, 182–189 (1985)

Ludwig, A., Schmeißer, C., Thänert, K.: Linear models in loss reserving. Dresdner Schriften zur Versicherungsmathematik 1/2009 (2009)

Ludwig, A., Schmeißer, C., Thänert, K.: Reserve risk estimation in a linear model. Z. Gesamte Versicherungswissenschaft **100**, 493–516 (2011)

Ludwig, A., Schmidt, K.D.: Calendar year reserves in the multivariate additive model. Dresdner Schriften zur Versicherungsmathematik 1/2010 (2010a)

Ludwig, A., Schmidt, K.D.: Gauss–Markov loss prediction in a linear model. Casualty Actuarial Society E-Forum Fall 2010, vol. 1 (2010b)

Mack, T.: Improved estimation of IBNR claims by credibility theory. Insur. Math. Econ. **9**, 51–57 (1990)

Mack, T.: A simple parametric model for rating automobile insurance or estimating IBNR claims reserves. ASTIN Bull. **21**, 93–103 (1991)

Mack, T.: Distribution-free calculation of the standard error of chain-ladder reserve estimates. ASTIN Bull. **23**, 213–225 (1993)

Mack, T.: Which stochastic model is underlying the chain-ladder method? Insur. Math. Econ. **15**, 133–138 (1994a)

Mack, T.: Measuring the variability of chain-ladder reserve estimates. In: Proceedings of the Casualty Actuarial Society Forum Spring 1994, vol. 1, pp. 101–182 (1994b)

Mack, T.: Credible claims reserves - the Benktander method. ASTIN Bull. **30**, 333–347 (2000)

Mack, T.: Schadenversicherungsmathematik, 2nd edn. Verlag Versicherungswirtschaft, Karlsruhe (2002)

Mack, T., Venter, G.: A comparison of stochastic models that reproduce chain-ladder reserve estimates. Insur. Math. Econ. **26**, 101–107 (2000)

Müller, P.H. (ed.): Wahrscheinlichkeitsrechnung und Mathematische Statistik - Lexikon der Stochastik, 5th edn. Akademie-Verlag, Berlin (1991)

Neuhaus, W.: Another pragmatic loss reserving method or Bornhuetter–Ferguson revisited. Scand. Actuar. J. **2**, 151–162 (1992)d

Panning, W.H.: Measuring loss reserve uncertainty. CAS Forum Fall **2006**, 237–267 (2006)

Pierson, F.: Using the whole triangle to estimate loss reserves. CAS Forum **1994**, 11–44 (1994)

Pröhl, C.. Schmidt, K.D.: Multivariate Chain-Ladder. Dresdner Schriften zur Versicherungsmathematik 3/2005 (2005)

Quarg, G., Mack, T.: Munich chain-ladder - a reserving method that reduces the gap between IBNR projections based on paid losses and IBNR projections based on incurred losses. Blätter DGVM **26**, 597–630 (2004)

Radtke, M.: Grundlagen der Kalkulation von Versicherungsprodukten in der Schaden- und Unfallversicherung. Verlag Versicherungswirtschaft, Karlsruhe (2008)

Reich, A., Zeller, W.: Spätschäden. In: *Handwörterbuch der Versicherung*, pp. 807–809. Verlag Versicherungswirtschaft, Karlsruhe (1988)

Schiegl, M.: On the safety loading of chain-ladder estimates - a Monte Carlo simulation study. ASTIN Bull. **32**, 107–128 (2002)

Schiegl, M.: A three dimensional stochastic model for claim reserving. ASTIN Colloquium Helsinki (2009)

Schiegl, M.: A model study about the applicability of the chain ladder method. Scand. Actuar. J. 482–499 (2015)

Schmidt, K.D.: Stochastische Modellierung in der Erfahrungstarifierung. Blätter DGVM **20**, 441–455 (1992)

Schmidt, K.D.: Lectures on Risk Theory. Teubner, Stuttgart (1996)

Schmidt, K.D.: Non-optimal prediction by the chain-ladder method. Insur. Math. Econ. **2**, 117–124 (1997)

Schmidt, K.D.: Prediction in the linear model - a direct approach. Metrika **48**, 141–147 (1998)

Schmidt, K.D.: Chain-ladder prediction and asset liability management. Blätter DGVM **24**, 1–9 (1999a)

Schmidt, K.D.: Reservierung für Spätschäden - Modellierung am Beispiel des Chain-Ladder Verfahrens. Allg. Statist. Archiv **83**, 267–280 (1999b)

Schmidt, K.D.: Loss prediction by generalized least squares - discussion of Halliwell (1996). Proc. CAS **86**, 736–747 (1999c)

Schmidt, K.D.: Methods and models of loss reserving based on run-off triangles - a unifying survey. CAS Forum Fall **2006**, 269–317 (2006)

Schmidt, K.D.: Optimal and additive loss reserving for dependent lines of business. CAS Forum Fall **2006**, 319–351 (2006)

Schmidt, K.D.: A note on the separation method. Dresdner Schriften zur Versicherungsmathematik 6/2007 (2007)

Schmidt, K.D.: Versicherungsmathematik, 3rd edn. Springer, Berlin (2009)

Schmidt, K.D.: Maß und Wahrscheinlichkeit, 2nd edn. Springer, Berlin (2011)

Schmidt, K.D.: Loss prediction based on run-off triangles. ASTA Adv. Stat. Anal. **96**, 265–310 (2012)

Schmidt, K.D., Schnaus, A.: An extension of Mack's model for the chain-ladder method. ASTIN Bull. **26**, 247–262 (1996)

Schmidt, K.D., Timpel, M.: Experience rating under weighted squared error loss. Blätter DGVM **22**, 289–307 (1995)

Schmidt, K.D., Wünsche, A.: Chain-ladder, marginal-sum and maximum- likelihood estimation. Blätter DGVM **23**, 267–277 (1998)

312

Schmidt, K.D., Zocher, M.: Loss reserving and Hofmann distributions. Mitt. SAV, 127–162 (2005)
Schmidt, K.D., Zocher, M.: The Bornhuetter-Ferguson principle. Variance **2**, 85–110 (2008)
Schweizer, R.: Spätschadenrückstellungen in der Rückversicherung. Schweizer Rück, Zürich (1989)
Sherman, R.E.: Extrapolating, smoothing and interpolating development factors. Proc. CAS **71**, 122–155 (1984)
Sherman, R.E.: Extrapolating, smoothing and interpolating development factors - author's reply. Proc. CAS **72**, 190–192 (1985)
Straub, E.: Non-life Insurance Mathematics. Springer, Berlin (1988)
Tarbell, T.F.: Incurred but not reported claim reserves. Proc. CAS **20**, 275–280 (1934)
Taylor, G.C.: Claims Reserving in Non-life Insurance. North-Holland, Amsterdam (1986)
Taylor, G.C.: Loss Reserving - An Actuarial Perspective. Kluwer, Boston (2000)
Van Eeghen, J.: Loss Reserving Methods. Nationale Nederlanden, Rotterdam (1981)
Verrall, R.J.: On the estimation of reserves from loglinear models. Insur. Math. Econ. **10**, 75–80 (1991)
Witting, T.: Kredibilitätsschätzungen für die Anzahl IBNR-Schäden. Blätter DGVM **18**, 45–58 (1987)
Wüthrich, M.V., Merz, M.: Stochastic Claims Reserving Methods in Insurance. Wiley, New York (2008)

List of Contributors

© Springer International Publishing Switzerland 2016
M. Radtke et al. (eds.), *Handbook on Loss Reserving*,
EAA Series, DOI 10.1007/978-3-319-30056-6

Author Index

© Springer International Publishing Switzerland 2016
M. Radtke et al. (eds.), *Handbook on Loss Reserving*,
EAA Series, DOI 10.1007/978-3-319-30056-6

Subject Index

Unbiased predictor, 65, 88, 93, 138, 140
Undertaking-specific parameters, 283
Underwriting year, 231, 242
Uniform distribution, 269, 302, 312
USP, 283

V
Value-at-risk, 265, 271, 281
Variance, 294, 295
Volume measure, 2, 13, 35, 43, 113, 125, 170, 192, 238, 244, 291, 307

W
Wald's identities, 76
Weibull curve, 287
Weighted squared approximation error, 135, 137, 147, 152

X
XL, 233

Y
Year, 247

Printed in the United States
by Baker & Taylor Publisher Services